Objective Tests and Revision Exercises

MATHEMATICS for SCHOOLS

Objective Tests and Revision Exercises

MATHEMATICS for Schools

LONDON

W. FOULSHAM & CO. LTD

NEW YORK · TORONTO · CAPE TOWN · SYDNEY

W. FOULSHAM & CO. LTD.,
Yeovil Road, Slough, Bucks., England

ISBN 0-572-00813-9
Made and Printed in Great Britain
by Pitman Press, Bath

Foreword

The need for general revision material, so necessary at the critical pre-examination stage, has led to a demand for a book such as this. Mainly written for the 'O' Level Certificate pupil, the easier examples provide ideal material for S.C.E., C.S.E. and similar courses.

Although written by the panel of authors responsible for the successful Mathematics for Schools — A Modern Certificate Course, the material is not specially referenced to that series and can therefore be used as ideal revision material with any series of modern mathematics text books.

The panel of authors by reason of their experience and in consideration of modern trends in examining, adjudged it wise to include a section of Objective Test Items — the remainder of the book consists of questions on a wide range of topics found in most mathematics syllabuses together with a short appendix on Polar coordinates.

The Objective Test Items include the main types usually found in most examination papers — multiple choice items, situation items and multiple completion items.

The revision exercises are classified by Topics and graded in order of difficulty.

There are over 350 original questions and exercises, and all material has been extensively tested in schools prior to publication.

Panel of teachers responsible for the preparation and writing

IAN D. WATT
Adviser in Mathematics, Glasgow Corporation Education Department

ALEXANDER A. CRAWFORD
Principal Teacher of Mathematics, King's Park Secondary School

SOLOMON FURST
Principal Teacher of Mathematics, Waverley Secondary School

PAUL SHERIDAN
Lecturer in Mathematics, Notre Dame College of Education

GEORGE B. P. SMITH
Principal Teacher of Mathematics, Knightswood Secondary School

GEORGE I. SUTHERLAND
Headmaster, Albert Secondary School

It is with pleasure that the members of the panel record their deep debt of gratitude to the Glasgow Corporation Education Committee, the Director of Education, John T. Bain, Esq., M.A., B.Sc., B.Ed., and Mathematics teachers in schools who, by their willing co-operation, made this venture possible.

Contents

Symbols

$=$	equals or is equal to
$\neq$	is not equal to
$\doteq$	is approximately equal to
$>$	is greater than
$\geqslant$	is greater than or equal to
$<$	is less than
$\leqslant$	is less than or equal to
$\{\ \ \}$	the set
$\{x:\ \}$	the set of replacements for x such that
$\in$	is a member of
$\notin$	is not a member of
$\subset$	is a subset of
$\cap$	intersection
$\cup$	union
$\{\ \}$ or $\emptyset$	empty set
E	Universal set
A'	the complement of the set A or the image of point A
$n(A)$	number of members of set A
N	the set of natural numbers, i.e. $\{1, 2, 3, \ldots\}$
W	the set of whole numbers, i.e. $\{0, 1, 2, 3, \ldots\}$
Z	the set of integers, i.e. $\{\ldots, -2, -1, 0, 1, 2, \ldots\}$
Q	the set of rational numbers, i.e. $\{x : x = \dfrac{p}{q}, p, q \in Z, q \neq 0\}$

9

R	the set of real numbers
R^+	the set of non-negative real numbers
$\rightarrow$	maps into, goes to, has as image
$f : x \rightarrow$	f maps x into
$f(x) =$	function f of x is equal to
$\Rightarrow$	implies or if . . . then . . .
$\Leftrightarrow$	has the same solution set as, is equivalent to (of statements), implies and is implied by
$\sqrt{}$	the square root of
$\sqrt[n]{}$	the nth root of
$\propto$	varies directly as
$\angle$	the angle
$\parallel$	is parallel to
$\perp$	is perpendicular to
$\triangle$	triangle or area of triangle
(x, y)	the ordered pair x, y
x_A	the x-coordinate of A
$\overrightarrow{AB}$	the displacement from A to B
$|\overrightarrow{AB}|$	the length of the displacement from A to B
$\simeq$	is equivalent to (of displacements)
$\underline{\mathbf{u}}$	the vector $\mathbf{u}$
$\begin{pmatrix} a \\ b \end{pmatrix}$	the translation or vector with x component, a, and y component, b
$\leftrightarrow$	$A \leftrightarrow B$, i.e. B is the image of A and A is the image of B
$[A, k]$	dilatation, centre A and scale factor, k
$A \circ B$	transformation B followed by transformation A or A after B

Part 1

Objective Test Questions

1 Multiple Choice Items

Algebra

(1) y varies as the square root of x, and k is a constant.
The value of y is:

A. $\dfrac{k}{x^2}$ B. $\dfrac{k}{\sqrt{x}}$ C. $k\sqrt{x}$ D. kx E. kx^2

(2) p, q and r are elements of a system which has two operations,
$*$ and $\circ$, such that $\circ$ is distributive over $*$. $p \circ (q * r)$ equals:

A. $(p \circ q) * (p \circ r)$ B. $(p * q) \circ (p * r)$ C. $p * (q \circ r)$
D. $(p \circ q) * r$ E. $(p * r) \circ p$

(3) $15 + 2x - x^2$ in product form is:

A. $(x - 5)(3 - x)$ B. $(5 - x)(3 - x)$ C. $(5 - x)(x - 3)$
D. $(x + 5)(x - 3)$ E. $(5 - x)(3 + x)$

(4) $4^2 \times 2^{-2}$ equals:

A. -64 B. -32 C. 1 D. 2 E. 4

(5) The solution set of the equation $(3x - 2)(2x + 3) = 0$ is:

A. $\left\{\dfrac{2}{3}, \dfrac{3}{2}\right\}$ B. $\left\{-\dfrac{2}{3}, \dfrac{3}{2}\right\}$ C. $\left\{-\dfrac{2}{3}, -\dfrac{3}{2}\right\}$

D. $\left\{-\dfrac{2}{3}, \dfrac{2}{3}\right\}$ E. $\left\{\dfrac{3}{2}, -\dfrac{3}{2}\right\}$

(6) In the Venn diagram, the shaded area represents:

A. $P \cap (Q \cup R)'$ B. $Q' \cap (P \cup R)$ C. $(P \cup R) \cap Q$
D. $(P \cap Q) \cup R$ E. None of these sets

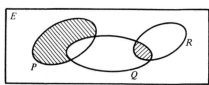

(7) $(4y^3)^2$ equals:

A. $16y^6$ B. $16y^5$ C. $8y^6$ D. $8y^5$ E. $6y^5$

(8) $\sqrt{\dfrac{4}{8}}$, when simplified, equals:

A. 2   B. $\sqrt{2}$ C. ½   D. $\dfrac{1}{\sqrt{2}}$ E. 1

(9) The value of $9^2 \times 3^{-4}$ is:

A. $-1\frac{1}{2}$ B. -1 C. 0   D. 1   E. None of these

(10) The coordinates of the maximum turning point of the curve
$y = 16 - x^2$ are:

A. (0, 16)   B. (0, 4)   C. (0, 0)   D. (4, 0)   E. (16, 0)

(11) The solution set of $(x - 1)(x - 2) = 2$ is:

A. $\{1, 2\}$ B. $\{0, 3\}$ C. $\{1\}$ D. $\{2\}$ E. None of these

(12) For real x and y, $(3x - 2y)^2$ equals:

A. $9x^2 - 12xy + 4y^2$ B. $9x^2 - 12xy - 4y^2$
C. $9x^2 - 6xy + 4y^2$ D. $9x^2 - 4y^2$
E. $9x^2 + 4y^2$

(13) Given that $2pq - p^2 = r^2$, the formula for q is:

A. $\dfrac{pr^2}{2}$ B. $r^2 + p^2 - 2p$ C. $\dfrac{r^2}{2p - p^2}$

D. $\dfrac{r^2 - p^2}{2p}$ E. $\dfrac{p^2 + r^2}{2p}$

(14) The region(s) representing $R \cap (P \cup Q)$:

A. 5   B. 2 and 5   C. 4, 5 and 6
D. 1, 2, 3, 4, 5, 6 and 7   E. None of these

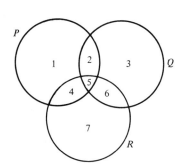

(15) The coordinates of the maximum turning point of the graph
$y = 6 - (x + 2)^2$ are:

A. $(-2, 0)$ B. $(-2, 6)$ C. $(0, -2)$ D. $(0, 2)$ E. $(6, -2)$

(16) If $a = \frac{1}{4}$, $a^{-\frac{1}{2}}$ has only one of the following values:

A. $\dfrac{1}{16}$ B. $\dfrac{1}{2}$ C. 2   D. 4   E. 16

(17) In the Venn diagram the shaded area represents one of these sets:

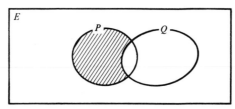

A. $P' \cup Q$ B. $P \cup Q'$ C. $P' \cap Q'$
D. $P' \cap Q$ E. $P \cap Q'$

(18) $3x - 2y = 7, x, \ y \in R$.
If $x = -1$, then y has one of these values:
A. 5   B. 2   C. -2 D. -5 E. -20

(19) If $f{:}x \rightarrow px^2 + qx + r$, then the roots of $f(x) = 0$ are:

A. $\dfrac{-p \pm \sqrt{p^2 - 4qr}}{2q}$ B. $\dfrac{-q \pm \sqrt{q^2 - 4pr}}{2p}$ C. $\dfrac{p \pm \sqrt{p^2 - 4qr}}{2p}$

D. $\dfrac{q \pm \sqrt{q^2 - 4pr}}{2p}$ E. $\dfrac{-q \pm \sqrt{p^2 - 4qr}}{2p}$

(20) A quadratic function, $y = f(x)$, is such that $f(b) = f(a) = 0$.
The equation of the axis of symmetry is:

A. $x = \dfrac{a - b}{2}$ B. $x = \dfrac{a + b}{2}$ C. $x = \dfrac{b - a}{2}$

D. $y = \dfrac{a - b}{2}$ E. $y = \dfrac{a + b}{2}$

(21) If $x \neq -3$ and $y \neq -3$, then
$$\frac{1}{x + 3} \times \frac{1}{y + 3} = \frac{1}{x + 3} + \frac{1}{y + 3}$$
is always true for:
A. $x + y = -5$ B. $x = y$ C. $x \times y = x + y$
D. No values of x and y E. All values of x and y except -3

(22) For $x \notin \{-5, -3\}$,
$$\frac{1}{x + 5} \times \frac{1}{x + 3} = \frac{1}{x + 5} + \frac{1}{x + 3}$$
is always true for:
A. $x = -4$
B. $x = -3\frac{1}{2}$
C. $(x + 5)(x + 3) = (x + 5) + (x + 3)$
D. No value of x
E. All values of x except -5 and -3

15

(23) The domain of the function of x in Fig. (i) is:

A. $\{1, 4\}$
B. $\{1, 2, 4\}$
C. $\{t: 1 \leqslant t \leqslant 4, t \in R\}$
D. $\{t: t \geqslant 1, t \in R\}$
E. R

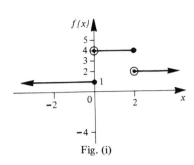

Fig. (i)

(24) $(27^{\frac{1}{3}})^{\frac{1}{2}}$ is equal to:

A. $\sqrt{3}$ B. 3   C. $27^{\frac{5}{6}}$ D. $\dfrac{9}{2}$ E. None of these

(25) X varies inversely as Y, and the constant of variation is 3. If X is doubled, then Y changes by the factor:

A. $\dfrac{1}{6}$ B. $\dfrac{1}{2}$ C. $\dfrac{2}{3}$ D. $\dfrac{3}{2}$ E. 6

(26) The range of this function of x in Fig. (ii) is:

A. $\{0, 2\}$
B. $\{0, 2, 4\}$
C. $\{t: 0 \leqslant t < 4, t \in R\}$
D. $\{t: t \geqslant 0, t \in R\}$
E. R

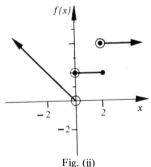

Fig. (ii)

(27) For $a = 17$ and $b = -13$, $a^2 - b^2$ is equal to:

A. 16   B. 30   C. 60   D. 120   E. 900

(28) $X^{\frac{1}{3}}(1 - X^{-\frac{1}{3}})$ is equal to:

A. $X^{\frac{1}{3}}$ B. $X^{\frac{2}{3}}$ C. $X^{\frac{1}{3}} + X^{\frac{1}{9}}$ D. $X^{\frac{1}{3}} - X^{-\frac{1}{9}}$ E. $X^{\frac{1}{3}} - 1$

(29) In the Venn diagram at the top of the next page $P \cup (Q \cap R)'$ is represented by regions:

A. 1
B. 1, 4 and 5
C. 1, 2, 3, 6 and 7
D. 1, 2, 3, 6, 7 and 8
E. 1, 2, 3, 4, 6, 7 and 8

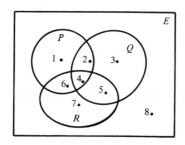

(30) If $x = 3a - 2b$ and $y = 4b - a$, then $2x - 3y$ is equal to:

A. $3a + 8b$ B. $3a - 16b$ C. $5a - 14b$

D. $9a + 8b$ E. $9a - 16b$

(31) The point (p, q) lies on the line $x - y = -6$. A pair of possible values for p and q is:

′ A. $(0, -6)$ B. $(6, -12)$ C. $(6, -6)$ D. $(6, .12)$ E. $(12, 6)$

(32) The value of $16^{\frac{3}{4}}$ is:

A. 12   B. 8   C. $\frac{1}{8}$ D. $-\frac{1}{8}$ E. -12

(33) For all values of x, $3x^3 . 2x^2$ equals:

A. $5x^5$ B. $5x^6$ C. $6x^5$ D. $6x^6$ E. $36x^5$

(34) The minimum value of the function $f(x) = (x - 2)^2 - 1$ is:

A. -5 B. -1 C. 0 D. 2 E. 3

(35) A quadratic function is such that $f(2) = f(-2)$. The equation of the axis of symmetry is:

A. $x = -2$ B. $x = 0$ C. $y = 0$ D. $x = 2$ E. $y = 2$

(36) The maximum value of the function f, where $f: x \rightarrow 9 - (x - 1)^2$, is:

A. 1   B. 8   C. 9   D. 10   E. None of these

(37) If $5 - 3x < 11$, the range of value of x is:

A. $x > -2$ B. $x < -2$ C. $x > 2$ D. $x < 2$ E. $-2 < x < 2$

(38) If $E = mc^2$, then the formula for c is:

A. $\sqrt{E - m}$ B. $\sqrt{E/m}$ C. $\sqrt{E/m}$ D. $\sqrt{mE}$ E. $\left(\frac{E}{m}\right)^2$

(39) When expressed as a single fraction, $\frac{1}{p} - \frac{1}{q}$ becomes:

A. $\frac{1}{p - q}$ B. $\frac{1}{pq}$ C. $\frac{p - q}{pq}$

D. $\frac{q - p}{pq}$ E. None of these

(40) The solution set of the equation $(x - 1)^2 = 1, x \in R$, is:

A. $\emptyset$ B. $\{0\}$ C. $\{0, 1\}$ D. $\{0, 2\}$ E. $\{2\}$

(41) $E = \dfrac{6\bar{a}^2}{(2\bar{a}^{-1})^3}$, when simplified and expressed with positive indices, becomes:

    A. $3a$ B. a C. $\frac{3}{4}a$ D. $\dfrac{a}{48}$ E. None of these

(42) Fig. (iii) shows a window composed of a square of side x, surmounted by a semi-circle. The perimeter of the window is:

    A. $x\left(4 + \dfrac{\pi}{2}\right)$ B. $x(3 + \pi)$

    C. $x\left(3 + \dfrac{\pi}{2}\right)$ D. $x^2\left(1 + \dfrac{\pi}{4}\right)$

    E. $x(3 + 2\pi)$

Fig. (iii)

(43) $a - 1 + \dfrac{1}{a+1} = \dfrac{P}{a+1}$, where P is:

    A. $a^2 - 2$ B. $a^2 - 2a$ C. $a^2 + 2$ D. a^2 E. a

(44) The identity element of this operation table is:

    A. k B. l C. m D. p E. q

$*$	k	l	m	p	q
k	k	l	m	p	q
l	l	m	p	q	k
m	m	p	q	k	l
p	p	q	k	l	m
q	q	k	l	m	p

(45) If $a*b$ means $a^2 - 2b^2$, then the value of $(-2)*3$ is:
    A. -32 B. -77 C. -16 D. -14 E. -8

Geometry

(1) If P' is the image of P under the dilatation $[Q, 2k]$, then $\overrightarrow{QP'}$ equals:

    A. $2k\overrightarrow{QP}$ B. $k\overrightarrow{QP}$ C. $\frac{1}{k}\overrightarrow{QP}$ D. $-k\overrightarrow{QP}$ E. $-2k\overrightarrow{QP}$

(2) A valid property of the diagonals of any kite is:
   A. They are equal in length.
   B. They bisect each other.
   C. They intersect each other at right angles.
   D. They bisect the angles of the kite.
   E. They are axes of symmetry of the kite.

(3) In Fig. (iv) $x^\circ + w^\circ$ equals:
   A. $x^\circ + y^\circ$ B. $x^\circ + z^\circ$ C. $y^\circ + z^\circ$ D. $360^\circ - (y^\circ + z^\circ)$
   E. 180°

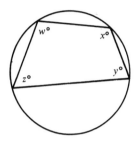

Fig. (iv)

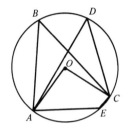

Fig. (v)

(4) In Fig. (v) $\angle ABC$ equals:
   A. $\angle BCD$ B. $\angle AEC$ (c) $2\angle AOC$ D. $\angle ADC$
   E. None of these

(5) $\underline{a} = \begin{pmatrix} 8 \\ -6 \end{pmatrix}$ and $\underline{b} = \begin{pmatrix} -4 \\ 3 \end{pmatrix}$, so $\underline{a}$ is equal to:

   A. $-2\underline{b}$ B. $-\frac{1}{2}\underline{b}$ C. $\frac{1}{2}\underline{b}$ D. $2\underline{b}$ E. None of these

(6) In Fig. (vi) $\underline{a} - \underline{b}$ is equal to:
   A. $\begin{pmatrix} -1 \\ 7 \end{pmatrix}$ B. $\begin{pmatrix} -1 \\ 1 \end{pmatrix}$ C. $\begin{pmatrix} 7 \\ 1 \end{pmatrix}$ D. $\begin{pmatrix} 1 \\ -7 \end{pmatrix}$ E. $\begin{pmatrix} 1 \\ -1 \end{pmatrix}$

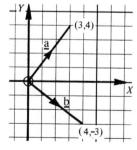

Fig. (vi)

19

(7) Fig. (vii) shows a cone whose base has a radius of 3 cm and whose height OC is 4 cm. The cone lies on a table and is free to roll round its apex O. As the cone rolls round O, the locus of C is a circle of radius:

A. $\dfrac{12}{5}$ cm   B. $\dfrac{16}{5}$ cm   C. 4 cm

D. 5 cm   E. None of these

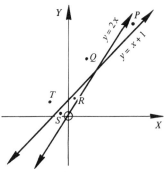

Fig. (vii)

(8) Fig. (viii) shows the graphs of $y = 2x$ and $y = x + 1$. The points P, Q, R, S and T are marked in certain regions. One of the following is false:

A. $y_P > x_P + 1$
B. $y_Q > 2x_Q$
C. $y_R > x_R + 1$
D. $y_S > 2x_S$
E. $y_T > x_T + 1$

Fig. (viii)

(9) Under reflection in the x-axis, $(3, 4)$ has image:
A. $(-4, 3)$ B. $(-3, -4)$ C. $(-3, 4)$ D. $(3, -4)$
E. $(4, -3)$

(10) In Fig. (ix) the size of $\angle\, QRS$ is:
A. $40°$ B. $60°$ C. $80°$ D. $100°$
E. Not determined from the information given

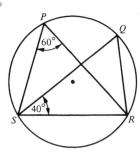

Fig. (ix)

(11) Under reflection in the line $y = 4$, (a, b) has image:
A. $(a - 8, b)$ B. $(8 - a, b)$ C. $(a, b - 8)$ D. $(a, 8 - b)$
E. None of these

(12) Under reflection in the line $x = a$, $(-a, b)$ has image:
A. $(0, b)$, B. $(a, -b)$ C. (a, b) D. $(2a, b)$ E. $(3a, b)$

(13) Reflection in $x = 2$ followed by reflection in $x = 1$ is equivalent to the translation:

A. $\begin{pmatrix} 1 \\ 0 \end{pmatrix}$ B. $\begin{pmatrix} 2 \\ 0 \end{pmatrix}$ C. $\begin{pmatrix} -1 \\ 0 \end{pmatrix}$ D. $\begin{pmatrix} -2 \\ 0 \end{pmatrix}$ E. $\begin{pmatrix} 0 \\ 2 \end{pmatrix}$

(14) If $\underline{a} = \begin{pmatrix} 3 \\ -1 \end{pmatrix}$ and $\underline{b} = \begin{pmatrix} -2 \\ 1 \end{pmatrix}$, then $3\underline{a} - 2\underline{b}$ equals:

A. $\begin{pmatrix} 5 \\ -5 \end{pmatrix}$ B. $\begin{pmatrix} 13 \\ -1 \end{pmatrix}$ C. $\begin{pmatrix} 13 \\ -5 \end{pmatrix}$

D. $\begin{pmatrix} 5 \\ -1 \end{pmatrix}$ E. $\begin{pmatrix} 13 \\ -2 \end{pmatrix}$

(15) In Fig. (x) $\overrightarrow{PQ}$ represents $\underline{a}$ and $\overrightarrow{QR}$ represents $\underline{b}$.
S divides PR in the ratio $3 : 2$. $\overrightarrow{SR}$ represents the vector:

A. $\frac{2}{3}(\underline{b} - \underline{a})$ B. $\frac{2}{5}(\underline{b} - \underline{a})$ C. $\frac{2}{3}(\underline{b} + \underline{a})$

D. $\frac{2}{5}(\underline{b} + \underline{a})$ E. $\frac{2}{5}(\underline{a} - \underline{b})$

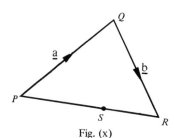

Fig. (x)

(16) The number of different types of quadrilaterals which have half-turn symmetry is:
A. 0   B. 1   C. 2   D. 3   E. 4

(17) H represents reflection in the line $x = -1$,
K represents reflection in the line $x = 2$,
P' is the image of $P(-3, 4)$ under the composite transformation $H \circ K$.
P' has coordinates:
A. $(3, 4)$ B. $(-3, 10)$ C. $(-9, 4)$ D. $(-3, -2)$ E. $(-5, 4)$

(18) The gradient of the line PQ in Fig. (xi) is:

A. $-\dfrac{6}{5}$ B. $-\dfrac{5}{6}$ C. $\dfrac{1}{2}$ D. $\dfrac{5}{6}$ E. $\dfrac{2}{1}$

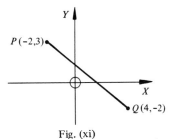

Fig. (xi)

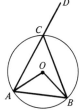

Fig. (xii)

(19) In Fig. (xii) O is the centre of the circle, and $\angle DCB = 130°$.
The size of $\angle OAB$ is:

A. $40°$ B. $50°$ C. $57·5°$ D. $65°$ E. $100°$

(20) $OP'Q'$ in Fig. (xiii) is the image of OPQ under a dilatation,
centre O. The scale factor is:

A. $-\dfrac{1}{3}$ B. $\dfrac{1}{3}$ C. 2   D. 3   E. -3

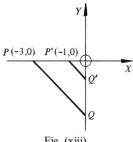

Fig. (xiii)

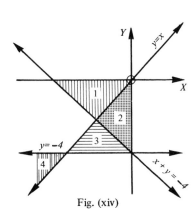

Fig. (xiv)

(21) In Fig. (xiv) the point $P(-3, -2)$ lies in the region:

A. 1   B. 2   C. 3   D. 4   E. None of these

(22) In Fig. (xv):

$\overrightarrow{OP}$ represents $\underset{\sim}{p}$, and $\overrightarrow{OQ}$ represents $\underset{\sim}{q}$

22

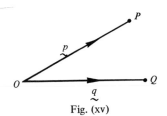

Fig. (xv)

$\overrightarrow{QP}$ represents:

A. $\underset{\sim}{q} - \underset{\sim}{p}$ B. $-(\underset{\sim}{p} + \underset{\sim}{q})$ C. $\underset{\sim}{p} - \underset{\sim}{q}$ D. $\underset{\sim}{p} + \underset{\sim}{q}$
E. None of these

(23) Under the transformation, reflection in $x = 7$ followed by reflection in $x = 4$, $P(-1, 3) \rightarrow P'$. The coordinates of P' are:
A. $(5, 3)$ B. $(-1, -3)$ C. $(-4, 3)$ D. $(-7, 3)$
E. None of these

(24) In Fig. (xvi) AB has the equation:

A. $3x - 4y - 9 = 0$
B. $4x - 3y - 12 = 0$
C. $3x + 4y + 6 = 0$
D. $4x + 3y + 12 = 0$
E. $4x + 3y - 12 = 0$

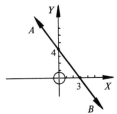

Fig. (xvi)

(25) If $\underline{u} = \begin{pmatrix} 3 \\ 2 \end{pmatrix}$ and $\underline{v} = \begin{pmatrix} 4 \\ -1 \end{pmatrix}$, then $3\underline{u} - 2\underline{v}$ is:

A. $\begin{pmatrix} 1 \\ 1 \end{pmatrix}$ B. $\begin{pmatrix} 17 \\ 4 \end{pmatrix}$ C. $\begin{pmatrix} 1 \\ 3 \end{pmatrix}$ D. $\begin{pmatrix} 13 \\ 4 \end{pmatrix}$ E. $\begin{pmatrix} 1 \\ 8 \end{pmatrix}$

(26) If $\underline{a} = \begin{pmatrix} -3 \\ 4 \end{pmatrix}$ and $\underline{b} = \begin{pmatrix} 12 \\ 5 \end{pmatrix}$, then the length of vector $(\underline{a} + \underline{b})$ is:
A. 9   B. $9\sqrt{2}$ C. 18   D. $\sqrt{82}$ E. None of these

(27) The image of $(-3, 2)$ under reflection in $(-1, 1)$ is:
A. $(-5, 3)$ B. $(1, 0)$ C. $(-1, 0)$ D. $(3, -2)$
E. $(5, -4)$

(28) Which shape has no axis of bilateral symmetry?
A. Parallelogram   B. Regular hexagon   C. Isosceles triangle
D. Rectangle   E. Rhombus

(29) The circle with centre $(0, 0)$ and radius 3 units has equation:
A. $x^2 - y^2 = 9$ B. $x^2 - y^2 = 3$ C. $x + y = 3$
D. $x^2 + y^2 = 3$ E. $x^2 + y^2 = 9$

(30) In Fig. (xvii) ∠ *PQR* equals:
   A. ∠ *RTS*
   B. ∠ *TSQ*
   C. ∠ *SQR*
   D. ∠ *QRT*
   E. None of these

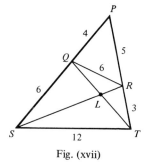

Fig. (xvii)

Trigonometry

(1) The minimum value of 6 cos 2*y* − 5 is:
   A. −17   B. −11   C. −5   D. 1   E. 7

(2) If cos *a* = cos 120° for 0° ⩽ *a* ⩽ 360°, the solution set is:
   A. {120°}   B. {300°}   C. {240°}   D. {120°, 300°}
   E. {120°, 240°}

(3) If *a* = 3 cos θ and *b* = 3 sin θ, *a*² + *b*² equals:
   A. 9 cos θ + 9 sin θ   B. ,9   C. 6   D. 3   E. None of these

(4) In Fig. (xviii) the value of *x* is:
   A. 5 sin 20°   B. 5 cos 20°   C. 5 tan 20°
   D. $\dfrac{5}{\sin 20°}$ E. $\dfrac{5}{\cos 20°}$

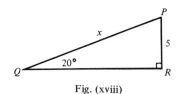

Fig. (xviii)

(5) The maximum value of 9 − 4 cos *X* is:
   A. 5   B. 7   C. 9   D. 11   E. 13

(6) The exact value of cos 30° is:
   A. $-\dfrac{1}{\sqrt{3}}$ B. $\dfrac{2}{\sqrt{3}}$ C. $\dfrac{\sqrt{3}}{2}$ D. $\sqrt{3}$ E. ½

24

(7) If in △ PQR, $\sin P = 0.8$, $\sin Q = 0.6$ and $p = 6$, then q has the value:

   A. $\dfrac{1}{8}$ B. $\dfrac{2}{9}$ C. $\dfrac{9}{2}$ D. 8

   E. Not determined from the information given

(8) Which of these answers give the exact value of $\cos 45°$?

   A. $\dfrac{1}{\sqrt{2}}$ B. 1   C. $\sqrt{2}$ D. $\dfrac{\sqrt{3}}{2}$ E. $\dfrac{1}{2}$

(9) In Fig. (xix) q is given by:

   A. $25 \sin 37°$ B. $\dfrac{25}{\sin 37°}$ C. $\dfrac{\sin 37°}{25}$ D. $25 \cos 37°$

   E. $\dfrac{25}{\cos 37°}$

Fig. (xix)

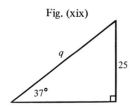

Fig. (xx)

(10) In Fig. (xx) p has the exact value:

   A. $6\sqrt{3}$ D. $\dfrac{24}{\sqrt{3}}$ C. 24   D. $24\sqrt{3}$ E. 6

(11) In Fig. (xxi) x has the value:

   A. $\dfrac{12 \sin 60°}{\sin 80°}$ B. $\dfrac{12 \sin 80}{\sin 60}$ C. $\dfrac{12 \sin 60°}{\sin 40°}$ D. $\dfrac{12 \sin 40°}{\sin 60°}$

   E. $\dfrac{12 \sin 80°}{\sin 40°}$

Fig. (xxi)

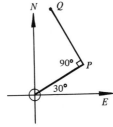

Fig. (xxii)

(12) In Fig. (xxii) the bearing of Q from P is:

   A. $030°$ B. $060°$ C. $240°$ D. $300°$ E. $330°$

25

(13) The function $2 \cos x - 1$, for $0° < x < 180°$, is always positive when:

A. $x < 60°$ B. $x > 60°$ C. $x > 120°$

D. $60° < x < 120°$ E. $x < 60°, x > 120°$

(14) The point P has coordinates $(-4, 3)$. Tan $\angle XOP$ equals:

A. $\dfrac{-4}{3}$ B. $\dfrac{-4}{5}$ C. $\dfrac{-3}{4}$ D. $\dfrac{3}{5}$ E. $\dfrac{3}{4}$

(15) The function $f(x) = 4 + 3 \sin x$ has a minimum value of:

A. -1 B. 0 C. 1 D. 4 E. 7

(16) In $\triangle ABC$, $AB = 2$ cm, $BC = 4$ cm and $CA = 3$ cm. Cos $\angle ACB$ is:

A. $\dfrac{1}{8}$ B. $\dfrac{11}{16}$ C. $\dfrac{3}{4}$ D. $\dfrac{7}{8}$ E. $-\dfrac{1}{4}$

(17) In $\triangle PQR$ in Fig. (xxiii) $PQ = 7$ units, $QR = 5$ units and $RP = 6$ units.

Sin $\angle PQR$ is equal to:

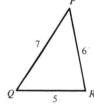

A. $\dfrac{7 \sin R}{6}$ B. $\dfrac{6 \sin R}{7}$

C. $\dfrac{6}{5 \sin P}$ D. $\dfrac{5 \sin P}{6}$

E. $\dfrac{7 \sin R}{5}$

Fig. (xxiii)

(18) Cos $225°$ has the value:

A. $-\dfrac{\sqrt{3}}{2}$ B. $\dfrac{\sqrt{3}}{2}$ C. $\dfrac{1}{\sqrt{3}}$ D. $\dfrac{1}{\sqrt{2}}$ E. $-\dfrac{1}{\sqrt{2}}$

(19) The period of $2 \sin x°$ is:

A. $90°$ B. $180°$ C. $270°$ D. $360°$ E. $720°$

(20) Which of the following graphs is most likely to be the graph of $\sin x°$ for $0 \leqslant x \leqslant 360$?

A.   B.

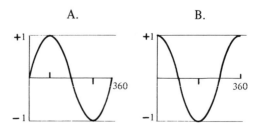

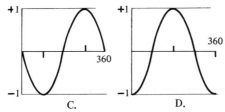

C.   D.

(21) The area of the triangle in Fig. (xxiv) in square units is:

     A. 24   B. $12\sqrt{3}$ C. 12   D. $6\sqrt{3}$ E. 6

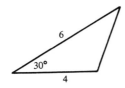

Fig. (xxiv)

(22) The latitude of A in Fig. (xxv) is measured by angle:

     A. *AOK*   B. *AON*   C. *KOL*   D. *GOL*   E. *AOG*

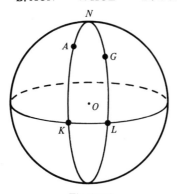

Fig. (xxv)

(23) The angle between *PM* and the plane *SRMN* in Fig. (xxvi) is:

     A. *PMN*   B. *PMS*   C. *SPM*   D. *MPR*   E. *PMR*

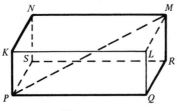

Fig. (xxvi)

27

2 Situation Items

(1) Each of the diagrams in Fig. (i), on an equilateral triangle tiling, shows a set of oblique axes OL and OM, the triangle shaded \\\\\\ and its image under a certain transformation shaded ▦

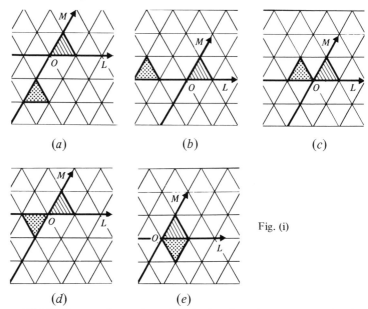

(a)   (b)   (c)

(d)   (e)   Fig. (i)

Which diagram shows the original triangle and its image under the transformation:

(i) Clockwise rotation of $60°$ about O

(ii) Translation $\begin{pmatrix} 0 \\ -2 \end{pmatrix}$ (iii) Half-turn about O (iv) Translation $\begin{pmatrix} -2 \\ 0 \end{pmatrix}$

(v) Anti-clockwise rotation of $120°$ about O

(2) Fig. (ii), which is not drawn to scale, shows part of the graphs of:
$y = x,\ x + y = 5,\ x + 5y = 10$

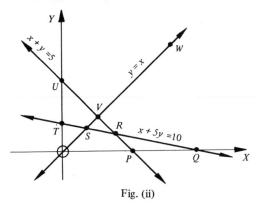

Fig. (ii)

(i) The coordinates of Q are:
   A $(5, 10)$ B. $(10, 0)$ C. $(0, 10)$ D. $(5, 5)$ E. $(2, 0)$

(ii) The coordinates of V are:
   A. $(2\frac{1}{2}, 2\frac{1}{2})$ B. $(5, 5)$ C. $(3, 2)$ D. $(2, 3)$
   E. None of these

(iii) The region which satisfies $x \geqslant 0$, $y \geqslant 0$ and
   $x + y \leqslant 5, x, y \in R$, is:
   A. TOQ B. VSR C. UTR D. $YUVW$
   E. None of these

(iv) The region which satisfies $x \geqslant 0$, $x + 5y \geqslant 10$ and
   $x + y \leqslant 5, x, y \in R$, is:
   A. TOQ B. VSR C. UTR D. UOV E. RPQ

(3) Fig. (iii) shows three narrow metal strips loosely jointed together
   to form an equilateral triangle ABC of side x units.

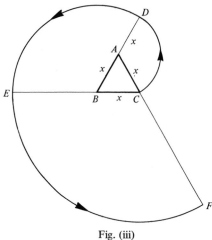

Fig. (iii)

29

The triangle is unwound so that AC is rotated about A into AD so that BAD is a straight line. Then BD is rotated about B into BE so that EBC is a straight line.

Finally EC is rotated about C until ACF is a straight line.

(i) The area of $\triangle ABC$ is:

A. $\frac{1}{2}x^2$ B. $\frac{3}{2}x^2$ C. $\frac{3}{4}x^2$

D. $\frac{1}{4}x^2$ E. None of these

(ii) The length of arc CD is:

A. $\frac{2\pi x}{3}$ units   B. $\frac{\pi x}{6}$ units   C. $\frac{\pi x^2}{3}$ units   D. $\frac{\pi x^2}{6}$ units

E. $\frac{2\pi x^2}{3}$ units

(iii) The area of sector ECF in square units is:

A. $9\pi x^2$ B. $3\pi x^2$ C. $6\pi x^2$ D. $\frac{3}{2}\pi x^2$

E. $2\pi x^2$

(4) In Fig. (iv) which graph is likely to be the graph of the function:

(i) $f(x) = (x - 3)^2$ (ii) $g(x) = x^2 - 2$ (iii) $h(x) = (x + 2)^2 + 2$

(iv) $k(x) = x^2$ (v) $l(x) = (x - 3)^2 - 2$

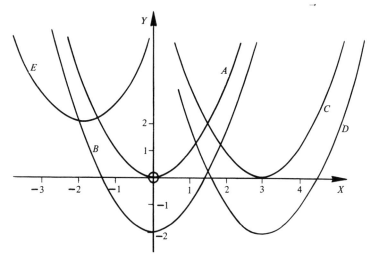

Fig. (iv)

3 Multiple Completion Items

Format I

In each of the following questions four statements (responses) –I, II, III and IV–are given.

Answer: A. if I, II and III only are correct
   B. if I and III only are correct
   C. if II and IV only are correct
   D. if IV only is correct
   E. if some other response or combination of the responses given is correct

Note. In each of the following quadrilaterals, questions (1)–(4), U, V, W and X are the mid-points of the respective sides.

(1) The following are axes of symmetry of the parallelogram *PQRS* in Fig. (i):

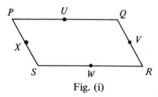

Fig. (i)

   I. *PR*   II. *XV*   III. *SQ*   IV. *UW*

(2) The following are axes of symmetry of the rectangle *PQRS* in Fig. (ii):
   I. *PR*   II. *XV*   III. *SQ*   IV. *UW*

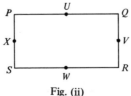

Fig. (ii)

(3) The following are axes of symmetry of the rhombus *PQRS* in Fig. (iii):

I. *PR*   II. *XV*   III. *SQ*   IV *UW*

Fig. (iii)

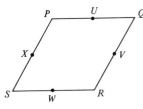

Fig. (iv)

(4) The following are axes of symmetry of the square *PQRS* in Fig. (iv):

I. *PR*   II. *XV*   III. *SQ*   IV. *UW*

(5) Under a rotation about any point *O*, which of the following are always true? A line segment *AB* and its image $A'B'$:

I. Are equal in length.

II. Intersect at *O* (*AB* and $A'B'$ produced if necessary).

III. Have the perpendicular bisectors of AA' and BB' concurrent at *O*.

IV. Have their perpendicular bisectors concurrent at *O*.

(6) In Fig. (v) $\angle ABC$ equals:

I. $\angle BCD$

II. $\angle AEC$

III. $\angle ADC$

IV. $\angle CEF$

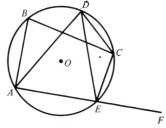

Fig. (v)

(7) Which of the following describes the dilatation [*Q*, 2], in which *R* is the image of *P*?

I. $\overrightarrow{QP} \simeq \overrightarrow{RQ}$ II. $\overrightarrow{QP} \simeq \overrightarrow{RP}$

III. $\overrightarrow{QP} \simeq 2\overrightarrow{QR}$ IV. $\overrightarrow{QR} \simeq 2\overrightarrow{QP}$

(8) Which of the following equals cos 40°?

I. cos 400°   II. sin 50°   III. sin 130°   IV. cos 140°

32

(9) Which of the following equations have the solution set $\varnothing$
$(0° \leqslant x \leqslant 180°)$?
    I. $2 \sin x = 3$ II. $4 \sin x = -3$ III. $\cos^2 x = -1$
    IV. $2 \tan x = 3$

(10) If $0 < X < 360$, $\tan X°$ is positive in:
    I. The first quadrant only   II. The third quadrant only
    III. The third and fourth   IV. The first and third
        quadrants   quadrants

(11) Sin 60° has the same value as:
    I. $\sin 120°$ II. $\cos 30°$ III. $-\cos 150°$ IV. $\sin 150°$

(12) Which of the following is true for Fig. (vi) in which
$\angle ABD = \theta°$ and $\angle DBX = \phi°$?
    I. $\sin \theta = \sin \phi$
    II. $\cos \theta - \cos \phi = 0$
    III. $\tan \theta + \tan \phi = 0$
    IV. $\tan \theta = \tan \phi$

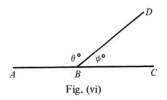

Fig. (vi)

(13) If $x = \sin \theta$, $y = \cos \theta$ and $z = \tan \theta$, which of the following
are true?
    I. $\dfrac{x}{y} = z$ II. $\dfrac{y}{x} = z$ III. $x^2 + y^2 = 1$ IV. $x^2 - y^2 = 1$

(14) Which of the following are equal to $\sin A$?
    I. $\cos (180-A)$ II. $\sin (180-A)$ III. $\sin (90-A)$
    IV. $\cos (90-A)$

(15) The universal set $E = \{1, 2, 3, 4\ 5, 6, 7, 8\}$,
and $H = \{1, 2, 4, 5, 6, 7, 8\}$. If $X \cap H = \{2\}$, then X is:
    I. $\{2\}$ II. $\{2, 3\}$ III. E IV. H'

(16) $F = \{1, 2, 3, 4, 5\}$, $G = \{3, 4, 5, 6, 7, 8\}$, and $H = \{1, 3, 5\}$, so
$(G \cap H) \cup F$ is a subset of:
    I. $\{2, 3, 4, 5, 6, 7, 8\}$ II. $\{3, 5\}$ III. $\{1, 3, 4, 5\}$ IV. $\{1, 2, 3, 4, 5\}$

(17) If $10 - x = 14 - y$, then:
    I. $y > x$ II. $y = x + 4$ III. $-x > -y$ IV. $-x > 4 - y$

(18) Which of the following are factors of $2x^2 - 7x - 15$:
    I. $2x - 3$ II. $x - 5$ III. $x + 5$ IV. $2x + 3$

(19) For all values of $a, b, c \in Z$, $(a * b) * c = a * (b * c)$, if $*$ is:
    I. $+$ II. $-$ III. $\times$ IV. $\div$

(20) The statement $a(b + c) = ab - ac$ is always true if:
    I. $a = 1$ II. $b = 0$ III. $b = -c$ IV. $c = 0$

33

(21) The solution set of the equation $\frac{x}{3} - \frac{x}{2} = \frac{1}{6}$ is:

    I. The additive inverse of 1

    II. The multiplicative inverse of -1

    III. A rational number

    IV. The identity element for multiplication

(22) $x^2 - 4y^2$ has as a factor:

    I. $x + 4y$ II. $x + 2y$ III. $x - 4y$ IV. $x - 2y$

(23) $x^2 + 9y^2$ has as a factor:

    I. $x - 3y$ II. $x + y$ III. $x + 3y$ IV. $x + 9y$

(24) The set of factors of $x^2 - x - 6$ contains:

    I. $x + 3$ II. $x + 2$ III. $x - 2$ IV. $x - 3$

Format II

In each of the following questions three statements (responses) are given.

Answer: A. if I, II and III are correct

          B. if I only is correct

          C. if II only is correct

          D. if III only is correct

          E. if I and II only are correct

(1) In Fig. (vii) the centre of rotation is P, and the angle of rotation is $180°$. A', B', C' and D' are the images of A, B, C and D. This rotation is equivalent to:

    I. Translation $\begin{pmatrix} -2 \\ 2 \end{pmatrix}$ II. Reflection in the line $y = x$

    III. Dilatation $[P, -1]$

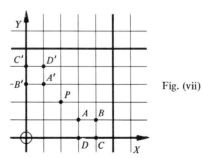

Fig. (vii)

34

(2) In Fig. (viii) $x° + z°$ equals:

I. $w° + y°$ II. $360° - (w° + y°)$

III. $360° - (x° + z°)$

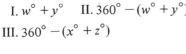

Fig. (viii)

(3) $\overrightarrow{AB}$ represents $\begin{pmatrix} 6 \\ 8 \end{pmatrix}$ and $\overrightarrow{BC}$ represents $\begin{pmatrix} 1½ \\ 2 \end{pmatrix}$.

I. $BC = 2½$ units   II. $\overrightarrow{AB} \simeq 4\overrightarrow{BC}$ III. A, B and C are collinear.

(4) Reflection in the line $y = x$ followed by reflection in the line $y = -x$ is equivalent to:

I. A rotation of $180°$ about the origin

II. The dilatation $[O, -1]$ III. Reflection in the origin

(5) The line $2y = 3x + 1$ is parallel to the line:

I. $2y - 3x = 0$ II. $3x - 2y + 5 = 0$ III. $2x - 3y = 0$

(6) Which of the following represent the sides of right-angled triangles?

I. $7, 24, 25$ units   II. $3, 4, 5$ units   III. $4, 5, 6$ units

(7) Which of the following does not represent the sides of a triangle?

I. $7, 5, 6$ units   II. $3, 8, 4$ units   III. $2, 5, 4$ units

(8) A member of the solution set of the equation $(y - 3)^2 = 1$ is:

I. 2   II. 3   III. -3

(9) The set $\{x: -2 \leqslant x < 1, x \in R\}$ has as a member:

I. 1   II. -3 III. $-½$

(10) Fig. (ix) shows part of the graph of a quadratic function, $y = f(x)$.

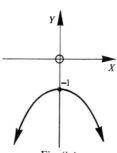

Fig. (ix)

I. The function has a maximum turning point.

II. The range is $\{x: x \leqslant -1, x \in R\}$.

III. The function has two real roots.

35

Part 2
General Revision
Questions

1 Algebra

Sets

(1) List the following sets:
   (a) Multiples of 8 which are less than 50.
   (b) The prime numbers between 30 and 40.
   (c) The months of the year which have 31 days.
   (d) $\{X : X = 2p + 1, p < 6, p \in N\}$.

(2) Which of the following are false?
   (a) $20 \in W$ (W is the set of whole numbers).
   (b) $3\frac{1}{2} \notin R$ (R is the set of real numbers).
   (c) The set of quadrilaterals with five interior angles is the empty set.
   (d) If $A = \{2, 4, 6\}$ and B is the set of the first three even numbers, then $A = B$.
   (e) The set of letters in the word LITTLE and the set of letters in the word TITLE are equal sets.
   (f) If A is the set of even numbers divisible by 3, then $A = \emptyset$.
   (g) If $P = \{a, b, c\}$, then P has exactly six possible subsets.

(3) $P = \{2, 4, 6, 8, 10\}, Q = \{3, 6, 9, 12, 15\}$ and $R = \{4, 10, 16, 22, 28\}$.
   (a) List the sets (i) $P \cap Q$, (ii) $P \cap R$, (iii) $Q \cup R$, (iv) $P \cup R$.
   (b) Find (i) $n(P)$, (ii) $n(Q \cup R)$, (iii) $n(P \cap R)$.

(4) (a) $A = \{p, q, r\}$ and $B = \{q, r, s\}$.   List $A \cup B$.
   (b) $P = \{1, 3, 5\}$ and $Q = \{2, 4, 6\}$.   List $P \cup Q$.
   (c) $X = \{\triangle, \square, \bigcirc\}$ and $Y = \{\triangle, \square\}$.   List $X \cup Y$.
   (d) $C = \{\times, +, -, \div\}$ and $D = \{-, \times, +, \div\}$. List $C \cup D$ and $C \cap D$.

(5) $E =$ the set of multiples of 3 less than 30, $A = \{12, 15, 18, 21, 24\}$ and $B = \{6, 9, 12, 15\}$.
   Draw a Venn diagram representing these sets.

(6) (a) Using the sets in (5) above, list:
   (i) A' (ii) B' (iii) $A \cap B'$ (iv) $A' \cup B$ (v) $A' \cap E$
   (vi) $A \cap E$

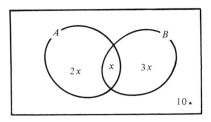

(b) The Venn diagram refers to a class of 40 pupils,
10 of which play neither badminton nor basketball.
A is the set of badminton players and
B is the set of basketball players.
$n(A) = 3x$, $n(B) = 4x$ and $n(A \cap B) = x$.
(i) Calculate the value of x.
(ii) How many pupils play badminton; basketball?

(7) In the following Venn diagrams:

(x)

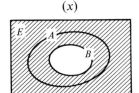

(y)

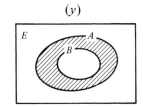

(z)

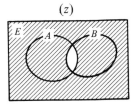

(a) Write the set represented by the shading in each case.
(b) (i) In which diagrams is $A \subset E$?
    (ii) In which diagrams is $A \subset B$?
    (iii) In which diagrams is $B \subset A$?

(8) In these Venn diagrams:

(i)

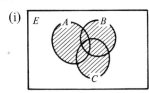

(ii)

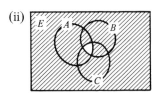

(iii) 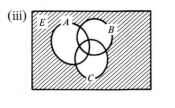 (iv)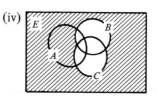

(v) 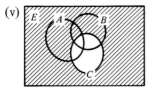 (vi)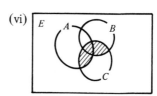

Write the sets represented by the shading in each case.

(9) E is the universal set, and A is any subset.

Using a Venn diagram if necessary, complete the following:

$E \cup A =$; $E \cap A =$; $A \cup A =$;

$A \cap A =$; $A \cup A' =$; $A \cap A' =$;

$E \cup A' =$; $E \cap A' =$

(10) On the following Venn diagrams, shade the sets indicated.

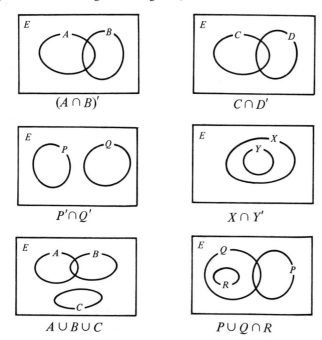

$(A \cap B)'$

$C \cap D'$

$P' \cap Q'$

$X \cap Y'$

$A \cup B \cup C$

$P \cup Q \cap R$

41

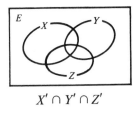

$X' \cap Y' \cap Z'$

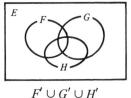

$F' \cup G' \cup H'$

(11) The Venn diagram represents the data about the viewing of 3
television programmes P, Q and R for a class of 21 pupils.

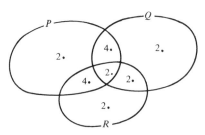

(a) How many watch programme P?
(b) How many watch programme Q?
(c) How many watch programme R?
(d) How many watch P and Q?
(e) How many watch Q and R?
(f) How many watch P and R?
(g) How many watch all three?
(h) Calculate how many view none of these programmes?
(i) How many view one programme only?

(12) The Venn diagram refers to a certain club where
P is the set of people who like walking;
Q is the set of people who prefer using buses; and
R is the set of people who have a bicycle.
Describe the sets represented by:
(a) shading ‖‖‖‖ (b) shading ≡ (c) shading ⬟⬟⬟

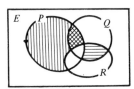

42

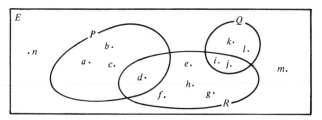

(13) For this Venn diagram fill in * in each of the following:
(a) $n(Q \cap R) = *$ (b) $P \cap * = \emptyset$ (c) $Q \cap R = \{i, *\}$
(d) $d \in (* \cap R)$ (e) $(P \cup Q \cup *)' = \{n, m\}$
(f) $P' \cap Q' \cap * = \{e, f, g, h\}$

Laws of Algebra

Commutative and Associative Laws

(1) Find the value of each of the following, if $a = 1, b = 2$ and $c = -3$:

(a) $a + 2b + c$ (b) $a^2 - 4ab$ (c) $a^2 - b^2 - c^2$

(d) $\dfrac{a^3}{3} - \dfrac{2b^3}{3} + 2c^3$ (e) $\dfrac{abc}{-3}$ (f) $\dfrac{ab + bc + ca}{abc}$

(g) $\dfrac{a}{b} + \dfrac{b}{c} + \dfrac{c}{a}$

(2) If $a * b = \dfrac{(-a) + 2b}{-3}$, find:

(a) $5 * 2$ (b) $2 * 5$ (c) $(-3) * 4$ (d) $4 * (-3)$
Is the operation * commutative?

(3) Construct an operation table for the set $\{-2, 2\}$ under the operation *, where * means "multiply and then divide by -2".

(4) Which of the following are true?
(a) $3(2 \times 1·75 - 2·5) \in W$.
(b) $(-6) \times (-2) - (5 \times (-3))$ is a positive integer.
(c) $\sqrt{196}$ is a rational number.
(d) $\sqrt{-¼} \in Q$.   (e) $\sqrt[3]{-648}$ is irrational.

(5) If $x * y$ means $5x + 5y$, evaluate:
(a) (i) $2 * 3$ (ii) $3 * 2$ (iii) $(-3) * 1$ (iv) $1 * (-3)$
   (v) $(3 * 4) * 2$ (vi) $3 * (4 * 2)$ (vii) $(a * 2a) * 3a$
   (viii) $a * (2a * 3a)$
(b) Is * (i) commutative or (ii) associative?
(c) If $x \oplus y$ means $bx + by$, is $\oplus$,
   (i) commutative or (ii) associative?

43

Distributive Law

(1) Use the distributive law to write the following as sums or differences of terms:

(a) $3(a + bc)$ (b) $4(1 - b)$ (c) $x(w - yz)$
(d) $-p(p + q)$ (e) $(x + y)w$ (f) $(4x - 3y^2)(-xy)$
(g) $5(p + qr)x$ (h) $-7(x - y)x$ (i) $3(a - b)(-b)^2$

(2) Use the distributive law to express, where possible, the following as products of factors:

(a) $3a + 3b$ (b) $a^2 + ab$ (c) $-xy - yz$
(d) $3x^2y + 6xy^2$ (e) $x^2 - 2xy - 3x^3$ (f) $a + bc + def$
(g) $-x^2 + x^2y$ (h) $(-x)^2 + x^2y$ (i) $(-xy)^2 - xy^2$

(3) Express as the product of factors, where possible:

(a) $3(2x + 3y) + 9(2x + y)$ (b) $2(3x + 4y) - 2(5x + 6y)$
(c) $x(9x - y) + x(x^3 + y)$ (d) $(x + y)^2 + (x + y)$
(e) $5(3x - 5y) - 4(4x - 6y) + x + y$
(f) $ax^2 - ay^2$
(g) $8x^2 - 2$ (h) $4(x^2 - 2y^2) - y^2$
(i) $x(9x - y) + y(x - y)$ (j) $(x + y)^3 - x - y$

(4) Express the following sums of terms as products of factors, where possible:

(a) $3a + 6b$ (b) $12p - 16q$ (c) $15 - 45ab$
(d) $ab - ac$ (e) $4xy - 6xz$ (f) $3a^2 - 9ab$
(g) $10pq - 5qr$ (h) $3x^2 - 6xy$ (i) $x^2y - xy^2$
(j) $7a + 14b - 21c$ (k) $a^2b + ab^2 - abc$ (l) $6 - 3ab + 12b$
(m) $a^2 - a$ (n) $a^2 + 4b$ (o) $2xy - 4x^2y$
(p) $6a^2b - 9bc$ (q) $6a^2b - 5cd$ (r) $2(x + y)^2 - 3(x + y)$

(5) Express the following products as sums or differences of terms:

(a) $(x + 3)(x + 2)$ (b) $(x + 5)(x + 7)$
(c) $(a - 4)(a - 5)$ (d) $(m - 7)(m - 1)$
(e) $(a + 3)(a - 1)$ (f) $(a + 3)(a - 5)$
(g) $(p + 10)(p - 11)$ (h) $(p + 7)(p - 7)$
(i) $(m - 3)(m + 3)$ (j) $(2x + 3)(3x + 2)$
(k) $(3x - 2)(2x - 3)$ (l) $(4x + 5)(5x - 4)$
(m) $(3x - 7)(4x + 1)$ (n) $(1 - 4x)(1 + 3x)$
(o) $(2x + 3y)(x - 2y)$ (p) $(3a + 2b)(2a - 3b)$
(q) $(a + 2b)(a + 2c)$ (r) $(3a + 2b)(3c + b)$
(s) $(a^2 - b^2)(7a^2 + 3b^2)$ (t) $(3x^2 - 2y)(3x^2 + 2y)$
(u) $(5a - 2b)(2b + 5a)$

(6) Expand the following:

(a) $(x + y)^2$ (b) $(3x + 2y)^2$ (c) $(5a + b)^2$
(d) $(p + 3q)^2$ (e) $(x - y)^2$ (f) $(2x - 3y)^2$
(g) $(4x - 3)^2$ (h) $(5 - x)^2$ (i) $(2 - 3x)^2$

(j) $(1 - 8x)^2$ $\quad$ (k) $(5ab - 3)^2$ $\quad$ (l) $(3a^2 + 2b)^2$

(m) $(6 + xy)^2$ $\quad$ (n) $(\pi - 2)^2$ $\quad$ (o) $(ac - 3b^2)^2$

(7) Using the expansion for $(a \pm b)^2$, evaluate:

(a) 31^2 $\quad$ (b) 29^2 $\quad\quad$ (c) 99^2 $\quad\quad$ (d) 101^2

(e) $1 \cdot 1^2$ $\quad$ (f) $1 \cdot 9^2$ $\quad\quad$ (g) $1 \cdot 01^2$ $\quad$ (h) $0 \cdot 99^2$

(8) Express in simplest form as sums or differences of terms:

(a) $3(x + 3y) + 4(2x - y)$

(b) $(x + 3)(x - 2) + (x + 5)(x + 7)$

(c) $(a - 4)(a - 5) - (a - 7)(a - 1)$

(d) $(3x - 7)(4x + 1) - (2x + 3)(3x - 2)$

(e) $2(3a + b)(a - 3b) - 3(2a - b)(a + 5b)$

(f) $(3x - 2)^2 + (4x - 3)^2$

(g) $2(3x - 2)^2 - 3(4x - 3)^2$

(h) $(5x + 3y)^2 - 3(2xy + y^2)$

(i) $(3 - 2p)^2 - 5(6 - 7p)$

(j) $4(3x - 4) - (2x + 3)^2$

(k) $(3x + 4)^2 + 7x + (2x + 3)(3x - 4)$

(l) $8 - (x + 3y)(4x - 5y)$

(m) $6xy - 2(x - y)(x - 2y)$

(n) $-(x - 3)^2 - (x - 2)(8 - x)$

(9) Factorise the following, where possible:

(a) $a^2 + 3a + 2$ $\qquad\qquad$ (b) $x^2 - 3x + 2$

(c) $p^2 - 3p - 4$ $\qquad\qquad$ (d) $m^2 + 3m - 4$

(e) $x^2 + 5x - 3$ $\qquad\qquad$ (f) $3a^2 - 2a - 1$

(g) $5m^2 + 6m - 8$ $\qquad\quad$ (h) $3x^2 + 10xy + 3y^2$

(i) $4x^2 - 17xy + 4y^2$ $\qquad$ (j) $6a^2 + 5a - 6$

(k) $6a^2 - 37a + 6$ $\qquad\quad$ (l) $12x^2 - 7xy - 12y^2$

(m) $12x^2 + 16xy - 35y^2$ $\quad$ (n) $12x^2 - 143xy - 12y^2$

(o) $12x^2 + 88xy - 15y^2$ $\quad$ (p) $24x^3 - 44x^2y - 40xy^2$

(q) $7a^3b^2 + 2a^4b + 6a^2b^3$ $\quad$ (r) $(a + b)^2 - 3(a + b) + 2$

(10) Factorise the following, where possible:

(a) $a^2 + 8a + 15$ $\qquad\qquad$ (b) $x^2 + 9x + 14$

(c) $a^2 + 3a - 18$ $\qquad\qquad$ (d) $m^2 + m - 30$

(e) $a^2 - 9a + 20$ $\qquad\qquad$ (f) $x^2 - 3x + 2$

(g) $p^2 + 3pq + 2q^2$ $\qquad\quad$ (h) $x^2 - 6xy + 7y^2$

(i) $x^2 - 8xy + 12y^2$ $\qquad\quad$ (j) $1 - 8y + 12y^2$

(k) $m^2 - 5mn + 4n^2$ $\qquad$ (l) $x^2 + 12x + 12$

(m) $12a^2 + 25a + 12$ $\qquad$ (n) $15x^2 + 11x - 12$

(o) $6x^2 + 7xy - 10y^2$ $\qquad$ (p) $15c^2 + 13cd - 20d^2$

(q) $10a^4 - a^2 - 3$ $\qquad\quad$ (r) $20 + 7a - 3a^2$

(s) $6 + 37p + 6p^2$ $\qquad\quad$ (t) $12a^2 - 16ab - 3b^2$

(u) $7x^2 + 3x + 2$ $\qquad\qquad$ (v) $16a^2 - 36a + 18$

(w) $18 - 51m + 36m^2$ $\qquad$ (x) $2x^4 + x^2y^2 - 3y^4$

(11) Find the factors of the following, where possible:
(a) $x^2 + 4xy + 4y^2$ (b) $6a^2 + 5ab + 6b^2$
(c) $12x^2 + 11x + 2$ (d) $15y^2 + 26y + 8$
(e) $6x^2 - 13xy + 6y^2$ (f) $12a^2 - 19ab - 21b^2$
(g) $24x^2 + 2x - 15$ (h) $12x^2 - 18x - 12$
(i) $12x^2y + 25xy + 12y$ (j) $30a^2 + 47a^2x + 14a^2x^2$
(k) $(x - 3)^2 + 5(x - 3) + 6$ (l) $(y + 2)^2 - (y + 2) - 6$

(12) Using "nesting", or otherwise, calculate $f(1), f(-1), f(2), f(-2)$, where:
(a) $f(x) = x^2 - 7x + 3$
(b) $f(x) = x^3 - 2x^2 + 4x - 1$
(c) $f(x) = 2x^4 + x^3 - 2x^2 + 3x + 1$

Difference of Squares

(1) Factorise the following:

(a) $7 - 7q^4$ (b) $3a^2b - 27b^3$ (c) $\dfrac{a^2}{4} - 4b^2$ (d) $a^2 - (b - c)^2$

(2) Find factors for the following, where possible:
(a) $x^2 - y^2$ (b) $1 - 4b^2$ (c) $9a^2 - 4b^2$
(b) $25a^2 - 49b^2$ (e) $121 - 36x^2$ (f) $100a^2 - 81c^2$
(g) $x^4 - y^4$ (h) $x^4y^2 - x^2y^4$ (i) $x^6y^2 - x^2y^6$
(j) $4x^8y^4 + 9x^4y^8$ (k) $16a^4 - 81b^4$ (l) $625 - x^4$
(m) $81x^2 + 36y^2$ (n) $a^3 - ab^2$ (o) $20x^2 - 5y^2$
(p) $x^3y^2 - x^2y^3$ (q) $63a^2 - 28b^2$ (r) $8x^3 - 18xy^4$

(3) Use factors to evaluate:
(a) $6 \cdot 5^2 - 5 \cdot 5^2$ (b) $4 \cdot 1^2 - 0 \cdot 9^2$ (c) $3 \cdot 4^2 - 1 \cdot 6^2$
(d) $9 \cdot 8^2 - 7 \cdot 2^2$ (e) $37^2 - 12^2$ (f) $2(8 \cdot 5)^2 - 2(8 \cdot 4)^2$

Miscellaneous

Find factors for the following, where possible:
(1) $3x^2 - 7x - 10$ (2) $xy^3 - x^3y$
(3) $2k + 18k^2$ (4) $15 - a(11 - 2a)$
(5) $10y^2 - 20y - 30$ (6) $3b^3 + 2b^2 - 1$
(7) $4(x + y)^2 - 1$ (8) $(a + b)^2 + 3(a + b)$
(9) $6z^2 - 5z - 56$ (10) $8\cos^2 A - 4\cos A$
(11) $3\tan^2 A - 4\tan A - 4$ (12) $5\sin^2 A - 20\cos^2 A$
(13) $3 + 11c - 4c^2$ (14) $(p + 2q)^2 - (p - 2q)^2$
(15) $p^4 - 5p^2 + 4$
(16) (a) $l^2 - 4m^2$
    (b) If $l^2 - 4m^2 = 24$ and $l - 2m = 4$, find the values of l and m

46

Algebraic Fractions

(1) Express in the simplest form:

(a) $\dfrac{5x}{x}$ (b) $\dfrac{10y^2}{4y}$ (c) $\dfrac{3a^2bc}{6ab}$ (d) $\dfrac{12p^3q}{9pq^3}$ (e) $\dfrac{21yz}{7xy^2z^2}$

(2) Find $*$ so that in each case the fractions are equivalent:

(a) $\dfrac{a}{8} = \dfrac{*}{8x}$ (b) $\dfrac{2y}{5x} = \dfrac{*}{10xy}$ (c) $\dfrac{3pq}{4x} = \dfrac{12pq^2}{*}$

(d) $\dfrac{25x^2y}{*} = \dfrac{5x}{4y}$

(3) Express as single fractions:

(a) (i) $\dfrac{p}{4} + \dfrac{p}{2}$ (ii) $\dfrac{2x}{3} = \dfrac{x}{6}$ (iii) $\dfrac{y}{5} - 2$ (iv) $\dfrac{3a}{4} + \dfrac{7a}{8} - a$

(b) (i) $\dfrac{2}{5} + \dfrac{1}{x}$ (ii) $\dfrac{3}{a} - \dfrac{1}{2a}$ (iii) $\dfrac{1}{2x} + \dfrac{1}{2y}$ (iv) $\dfrac{c}{d} + \dfrac{d}{c} + 1$

(v) $\dfrac{3x}{2y} + \dfrac{1}{4xy}$ (vi) $\dfrac{5}{2a} + ab$ (vii) $2p - 4 + \dfrac{1}{p}$

(viii) $\dfrac{a}{b} + \dfrac{b}{a} + 2$

(4) Express as a single fraction:

(a) $\dfrac{3}{x} + \dfrac{2}{x^2}$ (b) $\dfrac{3p}{2q^2} + \dfrac{5}{4q}$ (c) $\dfrac{a}{x} - \dfrac{3a^2}{x^3} + \dfrac{1}{x^2}$

(d) $\dfrac{2}{ab^2} + \dfrac{3}{a^2b} - \dfrac{4}{ab}$ (e) $\dfrac{3s}{2r^2} + \dfrac{5r}{2s^2} - \dfrac{1}{4}$

(5) Simplify the following:

(a) $\dfrac{4a^2b}{12ab^2}$ (b) $\dfrac{24x^8y^4}{2^3x^5y}$ (c) $\dfrac{5^23^3p^4q^5}{15p^6q^5}$ (d) $\dfrac{4(a+b)^2}{16(a+b)^5}$

(6) Simplify the following:

(a) $\dfrac{25xy}{-xy}$ (b) $\dfrac{-12a^2b}{-8ab}$ (c) $\dfrac{20(-1)^3m^2}{-41(-m)^2}$ (d) $\dfrac{(-y)^4\,(-z)^3x^2}{y^2z^4(-x)^4}$

(e) $\dfrac{(-1)^6(-2)^3k^2l}{4(-k)^4(-l)^2}$

(7) Using the distributive law, simplify:

(a) $\dfrac{3x + 6y}{3}$ (b) $\dfrac{5a^2 - 10a}{5a}$ (c) $\dfrac{12x^2y + 8xy^2}{4xy}$ (d) $\dfrac{6pq - 2q}{2q^2}$

(e) $\dfrac{2^3a^2b + (-2)^3ab}{4abc}$ (f) $\dfrac{2q^2 - 3q^3}{-q}$ (g) $\dfrac{x^4 - 3x^3 + x^2}{2(-x)^3}$

(8) Simplify the following:

(a) $\dfrac{a^2 - ab}{a^2 - b^2}$ (b) $\dfrac{a^2 + a - 20}{a^2 - 25}$

(c) $\dfrac{3a^2 + 15a + 18}{6a^2 + 36a + 48}$ (d) $\dfrac{x^2y}{x^2 - 9} \times \dfrac{x^2 + 5x + 6}{xy^2}$

47

$(e)\ \dfrac{x}{x+y} \times \dfrac{x^2 + 2xy + y^2}{x^2 - 3x}$

$(f)\ \dfrac{1}{a-b} \times \dfrac{a^2 - b^2}{2ab}$

$(g)\ \dfrac{7}{x} - \dfrac{3}{y}$

$(h)\ \dfrac{7}{xy} - \dfrac{3}{y}$

$(i)\ \dfrac{7}{x^2 y} - \dfrac{3}{xy^2}$

$(j)\ \dfrac{7}{x+y} - \dfrac{3}{x-y}$

$(k)\ \dfrac{7}{x-y} - \dfrac{3}{x+y}$

$(l)\ \dfrac{2}{a^2 - ab} + \dfrac{3}{ab - b^2}$

$(m)\ \dfrac{1}{x^2 - 9} - \dfrac{1}{x^2 - 5x + 6}$

$(n)\ \dfrac{1}{x^2 - 2xy + y^2} - \dfrac{1}{x^2 - y^2}$

(9) Simplify the following expressions:

$(a)\ \dfrac{x^2 - 36}{x^2 - x - 30}$

$(b)\ \dfrac{(ax + 4)^2}{8 - 2ax - a^2 x^2}$

$(c)\ \dfrac{10x^2 + 14x - 12}{(2 + x)(5x - 3)}$

$(d)\ \dfrac{a^2 - b^2}{a^2 + 2ab + b^2}$

(10) Simplify the following:

$(a)\ \dfrac{3x}{5} - \dfrac{2x}{7}$

$(b)\ \dfrac{3}{x+3} - \dfrac{2}{x-3}$

$(c)\ \dfrac{4}{x^2 - 3x} - \dfrac{5}{x-3}$

$(d)\ \dfrac{6}{x^2 + 7x + 12} - \dfrac{1}{x^2 + x - 12}$

(11) Simplify the following expressions:

$(a)\ \dfrac{3x^2 y^3}{6x^3 y}$

$(b)\ \dfrac{x^2 + 3x}{x^2 - 3x}$

$(c)\ \dfrac{x^2 + 4x + 4}{x^2 - 4}$

$(d)\ \dfrac{3x^2 + 9x + 6}{6x + 12}$

$(e)\ \dfrac{x^3 y - xy^3}{x^2 y + xy^2}$

$(f)\ \dfrac{2x^2 + 2y^2}{6x^4 - 6y^4}$

(12) $a^2 x + 2b^2 x = a^2 - 2ab + 3abx\ (a \neq 2b)$.

(a) Express x in terms of a and b as simply as possible.

(b) Find the value of x when $a = \frac{1}{2}$ and $b = \frac{1}{4}$.

(13) $x = a + \dfrac{2}{a}$ and $y = a - \dfrac{2}{a}$. Express the following in terms of a:

$(a)\ (x + y)^2$ $(b)\ (x - y)^2$ $(c)\ x^2 - y^2$ $(d)\ x^2 + y^2$

$(e)\ xy$ $(f)\ \dfrac{x}{y}$ $(g)\ \dfrac{x-y}{x+y}$

Equations and Inequations

Simple Equations

(1) Solve the following equations. In each case the variable belongs to the set of real numbers.

(a) $7x = 5x + 10$

(b) $8a + 4 = 4a + 12$

(c) $6x - 4 = 4x$

(d) $5b - 25 = 3b - 11$

(e) $2 + 3x = 6x - 7$ (f) $7 - 2x = 8x - 13$

(g) $5 - 9x = -4$ (h) $-17 = -5x - 4$

(i) $3 + 8x - 7 = 9x + 5 - 4x$ (j) $9a - 8 + 3a = 12 - 6a$

(k) $8 + 3x + 7 = 11 - x + 4$ (l) $6 - 3y = 8 - 2y$

(m) $3a - 15 + 5a - 15 = 2$

(n) $3m - 6 - 12m - 12 - 9m + 12 = 0$

(2) Form and solve an equation to find the answer to each of the following questions.

(a) x is a number. If I double it and add 4, the result is 14. What number does x represent?

(b) y is a number. If I multiply it by 3 and subtract 6, the result is 15. What number does y represent.

(c) One team scores 3 times as many goals as another. Although 12 goals are scored. What is the final score?

(d) On a train journey a child's fare is t pence, and an adult's fare is twice as much. The total fare for 2 adults and 3 children is £2·80. How much is a child's fare?

(3) Find the solution sets for the following equations, where the variable is a member of the set given:

(a) $3x = 12, \ x \in N$ (b) $4y + 1 = 12, \ y \in Q$

(c) $3y = 10, \ y \in W$ (d) $4x + 6 = 2 - 4x, \ x \in Q$

(e) $3x = 4x - x, \ x \in R$ (f) $3x = 4x + x, \ x \in R$

(g) $4x - 17 = 2(4 - 3x), \ x \in Q$

(h) $3(x + 2) = -4(x - 3), \ x \in Q$

(4) Solve the following equations. In each case the variable belongs to the set of real numbers.

(a) $4(x + 3) = 2(x - 3) + 10$

(b) $6(2a - 7) + 4 = 2(3a - 4)$

(c) $5(y + 2) - 46 + 9y = 7(y - 3) + 6$

(d) $5(a - 4) = 7 + 4(a + 3)$

(e) $6(2y - 3) + 5(2 - 3y) = 4(y + 2) + 5$

(f) $8(3x - 5) = 5(4x - 20) - 3(x + 1)$

(g) $5(3 - 2y) - 6(y - 1) = 7y - 3(y - 8) + 17$

(h) $3(2 - y) - 5(2y + 3) = 14 + 4(y + 7)$

(5) Form and solve an equation to find the answer to the following questions.

(a) x is a number. If I add 7 to it, and then multiply my answer by 5, I get the same answer as when I add one to it and then multiply by 11. What number does x represent?

(b) If I subtract 3 from a number, and then multiply my answer by 6, I get the same result as when I add one and then multiply by 4. What is the number?

(c) Adding 3 to a number, and then multiplying by 7 gives me one more than doubling the number, adding one and then multiplying by 5. What is the number?

(d) One rectangle has a length of x cm and a breadth of 3 cm. A second rectangle has length $3x$ cm and breadth 20 cm. The perimeter of the second rectangle is 4 times the perimeter of the first. What number does x represent?

(e) Owing to a rise in the price of apples of 4p per kg, I now pay 2p more for 3 kg than I used to pay for 4 kg. What was the original price per kg?

(f) As a result of a rise in the price of petrol of 1p per litre, a motorist receives only 18 litres for the price he used to pay for 20 litres. What is the new price per litre?

(6) Solve the following equations, where the variable is a member of the set of real numbers:

(a) $\dfrac{x}{3} + \dfrac{x}{4} = 14$

(b) $\dfrac{x}{7} - \dfrac{x}{6} = 3$

(c) $\dfrac{3x}{5} - \dfrac{2x}{3} = -3$

(d) $\dfrac{5x}{6} - \dfrac{7x}{4} = 2\tfrac{3}{4}$

(e) $\dfrac{2x-3}{10} + \dfrac{x+4}{5} = 2\tfrac{1}{2}$

(f) $\dfrac{3x-5}{4} = \dfrac{5x+10}{3}$

(g) $\dfrac{3(2y+3)}{5} = \dfrac{y-2}{2}$

(h) $\dfrac{3(3y-4)}{7} = \dfrac{2(2y+3)}{5}$

(i) $\dfrac{5(3y+1)}{12} = \dfrac{6y-11}{10}$

(j) $\dfrac{2}{3}(a-11) + \dfrac{3}{5}(a-5) + 3 = 0$

(k) $\dfrac{12}{7}(a-3) + \dfrac{1}{3}(a+5) = 2a - 3$

(l) $\dfrac{5}{6}(3a-2) - \dfrac{2}{3}\left(a - \dfrac{2}{3}\right) = 3a - \dfrac{1}{3}$

Simple Inequations

(1) Find the solution sets for the following inequations, where the variable is a member of the set given:

(a) $3x < 12, x \in N$

(b) $4y + 1 < 12, y \in \{0, 1, 2, 3\}$

(c) $3y + 1 \geqslant 12, y \in \{0, 1, 2, 3\}$

(d) $3x - 7 \leqslant 2 - 3x, x \in \{\tfrac{1}{2}, 1, 1\tfrac{1}{2}, 2, 2\tfrac{1}{2}\}$

(e) $2(y+3) \geqslant y + 5, y \in Z$

(f) $3x < 2x + x, x \in R$

(g) $-3(y-3) < 2(y+12), y \in Z$

(h) $4x \leqslant 2x + 2x, x \in R$

(2) Solve the following inequations, where $x \in R$:

(a) $6x - 9 \leqslant 15$

(b) $2x + 7 \geqslant 14$

(c) $18 - 5x < 8$

(d) $14 + 3x > 12$

(e) $3 - \frac{x}{2} < 8$

(f) $16 > 12 - \frac{x}{3}$

(g) $\frac{x}{2} - \frac{1}{6} < \frac{x}{3} - \frac{1}{8}$

(h) $\frac{5x - 1}{6} > \frac{5 + x}{9}$

(i) $\frac{2x + 3}{3} < \frac{3x - 4}{2}$

(j) $\frac{x}{4} + 7 \geqslant \frac{5 + x}{6}$

(k) $\frac{x - 4}{6} - \frac{5x}{12} \leqslant \frac{x - 2}{4}$

(l) $\frac{x - 3}{3} + \frac{1}{2} > \frac{x - 2}{2}$

(3) (a) A rectangle is $(x + 30)$ cm long and 5 cm broad. It is smaller than a square of side 15 cm. Find the solution set for x.

(b) The sides of a triangle have measures $3x, 4x - 5$ and $2x + 10$. Show that for this to be possible x must be greater than 3.

(c) A rectangle has length $44x$ cm and breadth $\frac{1}{14}(x + 20)$ cm. Its area is more than 88 cm² greater than a circle of radius x cm. Find the set of possible values of x. (Take $\pi = 3\frac{1}{7}$.)

Simultaneous Equations

(1) Solve the following equations, $x, y \in R$:

(a) $4x + 2y = 10$
$4x - 2y = 6$

(b) $3x + 5y = 14$
$6x - 5y = 13$

(c) $3x + 5y = 17$
$2x + 5y = 13$

(d) $8x + 5y = 22$
$8x - 3y = 38$

(e) $0 \cdot 5x + 0 \cdot 3y = -3$
$0 \cdot 4x + 0 \cdot 7y = 16$

(f) $7x + 3y = 26$
$5x - 2y = 60$

(g) $\frac{x}{3} + \frac{y}{4} = 4$
$0 \cdot 5x - 0 \cdot 2y = 1 \cdot 4$

(h) $\frac{3x}{5} - \frac{2y}{3} = 1$
$\frac{4x}{5} - y = 1$

(i) $\frac{2x}{3} + \frac{y}{5} = 3$
$\frac{1}{2}(y - x) = 1$

(j) $7x - y + 6 = 0$
$2x + 3y + 5 = 0$

(k) $14x + 13 = -3y$
$7 + 21x = -7y$

(l) $6x + \frac{1}{2} + 3y = 0$
$5 + 4y = 5x$

(2) (a) The length of a room is 2 m greater than its breadth, and its perimeter is 34 m. Find the length and breadth of the room.

51

(b) A sum of money has the value £2·65. It is made up from 10p and 5p pieces. In all there are 30 coins. How many of each kind are there?

(c) In Fig. (i) $\angle BAD = 50°$,
$\angle ABD = 55°$, $\angle ABC = x°$,
$\angle DBC = 2y°$, $\angle DCB = y°$

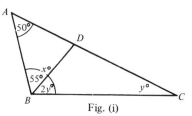

Fig. (i)

Complete the equations:
$x + y =$
$x - 2y =$
and hence find the replacements for x and y.

(d) The tens digit of a number exceeds the units digit by 5. The number is 5 less than 8 times the sum of its digits. What is the number?

(e) In a number in a base 6 system, the sixes digit exceeds the units digit by 2. The number is equal to 15_6 times its units digit. What is the number?

(f) The capacity of a school hall is 500. Seats at a school concert cost 15p and 25p. If $\frac{4}{5}$ of the dearer seats and $\frac{9}{10}$ of the cheaper seats were occupied, and the takings amounted to £77·25, how many seats of each kind were available?

Quadratic Equations

(1) Solve the following quadratic equations, where possible ($x \in R$ in each case), correct to one decimal place where necessary.
(a) $x^2 + x - 1 = 0$ (b) $x^2 - x + 1 = 0$ (c) $x^2 + 2x = 3$
(d) $2x^2 + 4x - 8 = 0$ (e) $2\frac{1}{2}x^2 = 5x + 7\frac{1}{2}$ (f) $2\frac{1}{2} = 1\frac{1}{4}x^2 - 3\frac{3}{4}x$
(g) $x^2 + \sqrt{13}x + 3 = 0$

(2) Solve the following quadratic equations, where $x \in R$. (Give solutions to two significant figures, if irrational.)
(a) $x^2 - 9x + 20 = 0$ (b) $x^2 + 11x = 26$ (c) $x^2 = 11x$
(d) $x^2 - 36 = 0$ (e) $2x^2 + 3x - 20 = 0$ (f) $2x^2 - x = 6$
(g) $6 = 6x^2 - 5x$ (h) $7x + 2 = 4x^2$ (i) $x^2 + 3x = 3$
(j) $x^2 + 3x - 5 = 0$ (k) $2x^2 + 8x + 3 = 0$ (l) $3x^2 + x = 5$

(3) Solve the following equations, where $x \in R$. Where the root is not exact, give the answer correct to two significant figures.

(a) $x^2 + 6x + 8 = 0$ (g) $x(x - 3) = 10$
(b) $x^2 - 3x = 5$ (h) $(x - 1)(x + 4) = 5$
(c) $3x^2 - 2x = 0$ (i) $5x = x^2$
(d) $5 - 4x - 3x^2 = 0$ (j) $3(x - 4) = (2x + 5)(4 - x)$

(e) $3x^2 - 13x + 4 = 0$ (k) $\dfrac{3}{x} = \dfrac{x + 1}{5}$

(f) $2x^2 - x - 2 = 0$ (l) $3x^2 = x + 7$

(4) A rectangle is such that the sum of its length and breadth is 7 m. Taking x m as the length of one side, find an expression for the area of the rectangle. Hence find the greatest value of the area of the rectangle and the length and breadth of the rectangle when the area has that value.

(5) $f(x) = x^2 - 4ax + (a + 3)^2 - 10$.
If $f(3) = 0$, find possible values of a.

(6) In Fig. (ii), which is not drawn to scale,
$ABCD$ is a rectangle. The lengths of AB, AD, AF, BG, CH and DE are $x + 2, x, 4, 3, 2$ and 3 cm respectively.

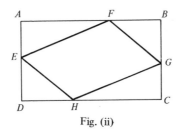

Fig. (ii)

(a) Calculate the area of each of the triangles AEF, FBG, GHC and EDH.
(b) Show that the area of quadrilateral $EFGH$ is $(x^2 - 5x + 18)$ cm^2.
(c) If the area of quadrilateral $EFGH$ is 24 cm^2, calculate the value of x.

(7) (a) (i) Simplify $\dfrac{x^2 - 2x}{x^2 + 2x} \times \dfrac{x^2 - 4}{x^2 - 4x + 4}$.

(ii) In (i) can $x \in \{-2, 0, 2\}$?

(b) Solve the equation $\dfrac{1}{y - 1} - \dfrac{1}{y + 1} = \dfrac{1}{4}$. $(y \in R, y \neq \pm 1)$

53

(8) (a) Find the solution set for $y(y + 3) = 40, y \in Z$.
(b) The length and breadth of a rectangle are $(2x + 5)$ cm and $(2x - 5)$ cm respectively. If the area is 75 cm^2, find the length and breadth of the rectangle.
(c) The length of a rectangle is $8x$ units and its breadth $6x$ units. A square is constructed with its side equal to the length of the diagonal of the rectangle. If the area of the square is 13 unit^2 greater than the area of the rectangle, calculate the value of x.

(9) A parabola whose equation is $y = x^2 + ax + b$ cuts the y-axis at C and the x-axis at A and B. C is $(0, -8)$, and A is $(2, 0)$.
(a) Find (i) the value of b; (ii) the value of a; and (iii) the co-ordinates of B.
(b) Deduce the equation of the axis of symmetry.
(c) Find the coordinates of the turning point.
(d) Sketch the graph of the function.

(10) 40 m of wire netting are used to enclose a rectangular piece of ground. If x m is the length of the rectangle, find:
(a) An expression for the area of the rectangle.
(b) The value of x for which the area is a maximum.
(c) This maximum area.

(11) The equation $3x = (2 + a)^2 + (a + 1)x$ is satisfied by $x = 2$. What are the two possible values of a?

Quadratic Inequations

(1) Solve the following quadratic inequations:
(a) $(x - 3)(x + 2) > 0$ (b) $x^2 + 2x - 3 \geqslant 0$
(c) $2x^2 - x - 6 < 0$ (d) $x^2 \geqslant 3x$

(2) Solve the following quadratic inequations using the graph of the function $f(x) = x^2 - 6x - 16$:
(a) $x^2 - 6x \geqslant 40$ (b) $x^2 - 6x \geqslant 27$ (c) $x^2 - 6x \leqslant 7$
(d) $x^2 - 6x \geqslant 0$ (e) $x^2 - 6x \geqslant -8$ (f) $x^2 - 6x \leqslant -5$

(3) A farmer has to build a rectangular cattle pen using a single strand of wire 400 m long. As the distance between the supporting posts has to be 5 m, the length of each side of the pen must be a multiple of 5 m. He also requires the area to be at least 9 900 m^2. Form an inequation, and hence calculate the possible lengths of the pen.

(4) Fig. (iii) shows a sketch of the graph of $y = 6x - x^2$, with the maximum turning point at C.

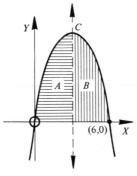

Fig. (iii)

(a) Find the coordinates of C.
(b) Using inequations, describe the regions:
    (i) A   (ii) B
(c) Draw in the graph of $y = 5$ and, using the shading ▦,
    show the region described by the following:
    $5 < y < 6x - x^2$

Relations, Mappings and Functions

(1) Each of the following sets of ordered pairs illustrates a relation.
For each set state if the relation is a function. If so, list the
domain and range and give the functional relationship, where
possible.
(a) $\{(1, 3), (-1, 1), (5, 7), (-3, -1)\}$
(b) $\{(1, 2), (2, 5), (3, 10), (-1, 2), (-2, 5)\}$
(c) $\{(-1, 0), (2, 3), (-1, -2), (-2, -3)\}$
(d) $\{(0, -5), (-1, -7), (-2, -9), (1, -3)\}$
(e) $\{(r, a), (s, b), (t, c), (u, d)\}$
(f) $\{(p, q), (r, s), (t, v), (r, w)\}$

(2) Find the maximum or minimum values (whichever is appropriate)
of the following functions and the corresponding value of the
independent variable.
(a) $f: x \rightarrow x^2 + 2x,\ x \in R$
(b) $g: y \rightarrow y^2 + 2y - 5,\ y \in R$
(c) $h: p \rightarrow 2p^2 + p - 1,\ p \in R$
(d) $f: x \rightarrow 10 - 6x - 6x^2,\ x \in R$
(e) $g: r \rightarrow 3r^2 + 2r + 6,\ x \in R$

55

(3) Show on separate Cartesian diagrams sketches of each of the following functions, indicating clearly the turning point and, where possible, intersections with the axis.

(a) $f: x \to x^2 - 3x$, $x \in R$ (b) $g: x \to x^2 - 3x + 5$, $x \in R$

(c) $f: x \to 2x^2 + 9x - 5$, $x \in R$ (d) $g: x \to 15 - 2x - x^2$, $x \in R$

(4) Illustrate by an arrow diagram the relation $\Rightarrow$ on the set $\{a, \beta, \theta, \phi\}$, where a, β, θ and ϕ represent respectively the statements:

$x = 0$, $x^2 - x = 0$, $(x - 1) = 0$ and $x^2 = x$.

(5) The following set of ordered pairs represent mappings of a set A into a set B. In each case what are sets A and B?

(a) $\{(1, 2), (2, 3), (3, 4), (4, 5), (5, 6)\}$

(b) $\{(1, 7), (2, 6), (3, 5), (4, 4), (5, 3), (6, 2), (7, 1)\}$

(c) $\{(x, y): y = 2x, x \in \{1, 2, 3, 4, 5\}\}$

(d) $\{(x, y): y = x^2 + 1, -1 \leqslant x \leqslant 1, x \in R\}$

(6) For the following sets of arrow diagrams and ordered pairs:

(i) Which set of ordered pairs corresponds to which diagram?

(ii) Which diagrams represent mappings of A into B?

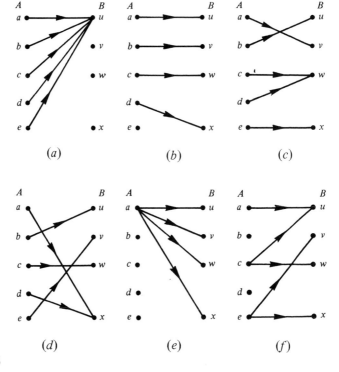

(a)   (b)   (c)

(d)   (e)   (f)

(g) $\{(a, x), (b, u), (c, w), (d, x), (e, v)\}$
(h) $\{(a, u), (b, v), (c, w), (d, x)\}$
(i) $\{(a, u), (b, u), (c, u), (d, u), (e, u)\}$
(j) $\{(a, u), (a, v), (a, w), (a, x)\}$
(k) $\{(a, u), (c, u), (c, w), (e, v), (e, x)\}$
(l) $\{(a, v), (b, u), (c, w), (d, w), (e, x)\}$

(7) f is the function defined by $f(x) = x^2 - 3x - 10$.
 (a) Find the solution set of $f(x) = 0$.
 (b) Find the coordinates of the points where the graph of f cuts
  the x- and y-axes.
 (c) Write down the equation of the axis of symmetry.
 (d) Sketch the graph of f and its axis of symmetry.
 (e) Calculate the minimum value of $f(x)$.
 (f) What is the minimum value of $h(x) = x^2 - 3x - 12$?
 (g) What is the maximum value of $g(x) = 12 + 3x - x^2$.

(8) Fig. (iv) shows the graph of the mapping
 $f(x) = x^3 - 2x^2 - x + c, \ x \in R$.
 (a) State the value of c.
 (b) If $f(p) = 0$, state the values of p.
 (c) State the value of $f(1)$.
 (d) How many values of x are mapped into 5?
 (e) For what values of a is $f(a) = f(1)$?

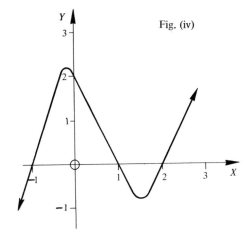

Fig. (iv)

(9) (a) Express the mapping
  $x \to x^2 + x - 2, x \in \{-3, -2, -1, 0, 1, 2, 3\}$ by:
   (i) An arrow diagram (ii) A set of ordered pairs
   (iii) A graph
 (b) Write down the solution set of the equation $x^2 + x - 2 = 0$.

(10) (a) A and B are two sets. $n(A) = 2, n(A) = 3$. How many mappings are there from A into B?

(b) The diagram shows a game in which the competitors are asked to roll balls A, B and C into the slots 1, 2, 3, 4 or 5. In how may ways can this be done? (Remember, more than one ball may end up in the same slot.)

(11) A firm's price code consists of a 10 letter word which has a one-to-one correspondence with the 10 numbers 0 to 9. If £UN.DC represents £18·90 and £MR·BL represents £25·36, deduce the codeword by finding the remaining two letters. Write the code prices for £7·55 and £0·49.

(12) A function $g(x)$ is defined for a domain $\{0, 1, 2, 3, 4\}$ by $g(x) = 1$ for $0 \leqslant x \leqslant 2$ and $g(x) = x - 1$ for $2 < x \leqslant 4$. Find the range of x and illustrate g on an arrow diagram.

(13) $P = \{x < 14: x - 1$ is a prime number$\}$
(a) List the members of P.
(b) A function f is defined on P as follows:

$$x \to \frac{x}{2}, x \text{ even} \qquad x \to \frac{x+1}{2}, x \text{ odd}$$

State the range of f and illustrate f by means of an arrow diagram.

(14) By completing the square, find the coordinates of the turning points of the graphs of the following quadratic functions, where $x \in R$. In each case state the equation of the axis of symmetry and whether the turning point is a maximum or a minimum.

(a) $f(x) = x^2 + 4x + 3$ (e) $f(x) = 5x + x^2 + 3$
(b) $f(x) = x^2 + 6x + 12$ (f) $f(x) = 8 - 3x - x^2$
(c) $f(x) = -x^2 - 4x + 8$ (g) $f(x) = 2x^2 + 9x + 6$
(d) $f(x) = -x^2 - 6x + 2$ (h) $f(x) = 8 - x - 3x^2$

(15) The skeleton of a box with a square base has to be constructed from pieces of wood of total length 24 m (see Fig. (v)).

58

Taking the length of each side of the base to be
x m, show that:

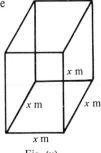

(a) The height of the box is $6 - 2x$ m.
(b) The total surface area of the box
will be $(24x - 6x^2)$ m².
(c) This surface area is a maximum when
the box has the shape of a cube.

Fig. (v)

(16) A function $f(x)$ is defined by $f(x) = 2x^2 - 1, x \in R$.
(a) Find the image of -2.
(b) Find the elements of the domain with image of $3·5$.
(c) If $f(m) = 49$, find m.
(d) Show graphically or otherwise that no member of the range
is less than -1.

(17) Which of the following are true for $f(x) = x^2 - 4x + 9, x \in R$?
(a) $f(0) = 9$.
(b) The minimum turning value is 5.
(c) The range is $\{p : p \leqslant 5\}$.
(d) When $f(b)$ is 9, b has the values 0 and 4.

(18) Fig. (vi) shows a sketch of the curve $y = 25 - x^2$. PQ is perpendicular to OX.

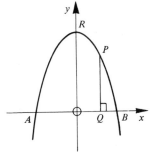

Fig. (vi)

(a) If $OQ = 3$ units, find the coordinates of P.
(b) $S\,(a, b)$ is a point on the curve. Write down the coordinates
of S', the image of S under the reflection in the axis of
symmetry of the curve.

59

(c) Find the solution set of the system of equations $y = 0$,
   $y = 25 - x^2$.
(d) Write down the coordinates of the turning point R.
(e) For what subset of the domain is $25 - x^2$ positive?

(19) Fig. (vii) shows the graph of $f(x) = x^2 - 4x$.

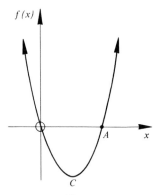

Fig. (vii)

(a) Find the coordinates of A.
(b) Write down the equation of the axis of symmetry.
(c) Write down the x-coordinate of the turning point C.
(d) Evaluate $f(2)$.
(e) Write down the coordinates of C.
(f) Give the range of values of x for which $f(x) < 0$.
(g) Does $(8,32)$ lie on the graph?
(h) If $(p,5)$ lies on the graph, calculate the possible values of p.
(i) $D(-2,k)$ is a point on the graph. Find the value of k.
(j) Find the image of D under reflection in the axis of symmetry of the graph.
(k) Draw a sketch of the graph under reflection in the y-axis.
(l) Draw a sketch of the graph under reflection in the x-axis.

(20) A function f is defined by $f: x \to x^2 + x - 6$.
   (a) Evaluate $f(4)$.   (b) Evaluate $f(2^{-1})$.
   (c) At what points does the graph of f cross the x-axis?
   (d) At what point does the graph of f cross the y-axis?
   (e) $f(a) = -4$. What are the values of a?

Linear Programming

(1) A factory worker takes 1½ hours and 1 kg of raw material to make product X, and ½ hour and 1 kg of raw material to make product Y. Not more than 12 kg of raw material have to be used each day, and the worker works a maximum of 9 hours a day. Product X produces a profit of £4 and product Y a profit of £2 on sale.
How many of each article should be produced each day for maximum profit? What is the amount of this profit?

(2) A firm produces two types of machine for export. Machine A weighs 2 tonnes and occupies $10\,m^3$. Machine B weighs 1 tonne and occupies $20\,m^3$. Shipping space is limited to $60\,000\,m^3$, and the total weight must not exceed 6 000 tonnes. Each machine A brings in a profit of £50 and each machine B a profit of £40. How many of each should be exported for maximum profit? What is the amount of this profit?

(3) Machine A can produce articles X at the rate of 100/hour. Machine B can produce articles Y at the rate of 150/hour. Not more than 900 articles are to be produced each day, and the total daily running time of the machines must not exceed 7 hours. Each article X makes a profit on sale of 50p, and each article Y makes a profit on sale of 40p.
(a) For how many hours should each machine be run in order to obtain maximum profit, and what is the value of this profit?
(b) Later, owing to maintenance difficulties, machine A can run for a maximum of 5 hours a day. At the same time market changes alter the price structure so that each article Y makes a profit of 30p and each article X a profit of 60p on sale. For how many hours should each machine now be run, and what is the amount of the new profit?

(4) Two types of wagons are used in a goods yard. Wagon A carries 20 tonnes and is 10 m in length. Wagon B carries 10 tonnes and is 7 m in length. The total length of wagons which can be accommodated at any one time is 378 m. Owing to handling difficulties, the number of wagons A must not be more than twice the number of wagons B.
How many of each type should be selected in order to achieve the maximum weight of goods loaded? What is the value of this maximum weight?

(5) In a supermarket display stands of floor area 7 m² and 4 m² are to be used. The total floor area available for these stands is 308 m². Owing to the shape of the supermarket, the number of

large stands must not be more than the number of small stands. The average daily profit from the small stands is £30 and from the large stands £70.

How many of each stand should be used for maximum profit, and what is this profit?

Surds

(1) Simplify the following:

(a) $\sqrt{8}$ (b) $\sqrt{72}$ (c) $\sqrt{75}$

(d) $\sqrt{405}$ (e) $\sqrt{32} \times \sqrt{8}$ (f) $\sqrt{27} \times \sqrt{3}$

(g) $\sqrt{125} \times \sqrt{5}$ (h) $\sqrt{6} \times \sqrt{15}$ (i) $\sqrt{14} \times \sqrt{35}$

(j) $\sqrt{243} \div \sqrt{27}$ (k) $\sqrt{3\,125} \div \sqrt{125}$ (l) $\sqrt{28} \div \sqrt{12}$

(m) $2\sqrt{3} \times 3\sqrt{27}$ (n) $4\sqrt{3} \times 2\sqrt{12}$ (o) $3\sqrt{3} \times 5\sqrt{6}$

(p) $4\sqrt{2} \times 2\sqrt{14}$ (q) $6\sqrt{18} \div 3\sqrt{8}$ (r) $5\sqrt{21} \div 4\sqrt{28}$

(s) $\sqrt{8} \div 2\sqrt{2}$

(t) $\sqrt{8} + \sqrt{32} + \sqrt{128}$

(u) $\sqrt{3} - 3\sqrt{27} + 5\sqrt{243}$

(v) $6\sqrt{3} - 3\sqrt{5} - 2\sqrt{27} + 4\sqrt{125}$

(2) Express with rational denominators:

(a) $\dfrac{\sqrt{2}}{\sqrt{3}}$ (b) $\dfrac{4\sqrt{3}}{\sqrt{5}}$ (c) $\dfrac{2\sqrt{15}}{\sqrt{5}}$ (d) $\dfrac{3\sqrt{6}}{5\sqrt{12}}$

(3) Simplify the following:

(a) $\sqrt[3]{81}$ (b) $\sqrt[3]{16} \times \sqrt[3]{4}$ (c) $4\sqrt[3]{32} \times 3\sqrt[3]{4}$

(d) $\sqrt[3]{625} \div \sqrt[3]{5}$ (e) $6\sqrt[3]{2} - 3\sqrt[3]{3} + 2\sqrt[3]{16} + 5\sqrt[3]{81}$

(f) $\sqrt[3]{3} + \sqrt[3]{81} + \sqrt[3]{2\,187}$

Indices

(1) Simplify the following, where possible:

(a) $y \times y \times y \times y$ (b) $C^3 \times C^4$ (c) $a^7 \times a$

(d) $b^6 \div b^2$ (e) $d^5 \div d^5$ (f) $e^4 \div e^7$

(g) $(12x)^2 \div 12x^2$ (h) $6y^3 \times (2y)^3$ (i) $(0 \cdot 1x)^3 \div 0 \cdot 1$

(j) $3a^2 \times 2b^3$ (k) $1 \div 5a^6$ (l) $a^6 \div \dfrac{1}{a^7}$

(2) Simplify the following, where possible:

(a) $a^4 \times a^{-3}$ (b) $y^7 \div y^{-2}$ (c) $x^\circ \div x^{-4}$

(d) $a^{-7} \div a^{-5}$ (e) $8y^3 \times (2y)^{-3}$ (f) $7y^{-7} \times 8z^{-8}$

(g) $5a^{-4} \div 1$ (h) $\dfrac{1}{x^{-2}y^{-3}}$

(3) Evaluate the following, when $x = 8$, $y = 9$:

(a) x° (b) $y^{\frac{1}{2}}$ (c) $x^{\frac{1}{3}}$ (d) y^{-2}

(e) $x^{-\frac{2}{3}}$ (f) $y^{-\frac{3}{2}}$ (g) $4y^{\frac{1}{2}}$ (h) $(4y)^{\frac{1}{2}}$

(i) $27x^{\frac{1}{3}}$ (j) $(27x)^{-\frac{1}{3}}$ (k) $\frac{1}{2}xy^{\frac{3}{2}}$ (l) $\left(\frac{1}{2}xy\right)^{\frac{3}{2}}$

(m) $(x^2 y)^{\frac{1}{2}}$ (n) $(3xy)^{-\frac{1}{3}}$ (o) $\left(\dfrac{x}{y^2}\right)^{-\frac{4}{3}}$ (p) $\left(\dfrac{x^2}{y}\right)^{-\frac{3}{2}}$

(q) $\left(\dfrac{4x}{3y^2}\right)^{-\frac{1}{5}}$

(4) Find solution sets for the following, where $x \in Q$:

(a) $x^{\frac{1}{2}} = 4$ (b) $4x^{\frac{1}{3}} = 8$ (c) $x^{-2} = \frac{1}{4}$ (d) $8x^\circ = 8$

(e) $\frac{1}{2}x^\circ = 1$ (f) $x^{\frac{3}{4}} = 27$ (g) $9x^{\frac{1}{2}} = 18$ (h) $(9x)^{\frac{1}{2}} = 18$

(i) $3x^{\frac{5}{3}} = 96$ (j) $4^x = 8$ (k) $9^x = 27$ (l) $2^x = 16$

(m) $2^x = -16$ (n) $3^x = \dfrac{1}{27}$ (o) $25.5^x = 1$ (p) $(25.5)^x = 1$

(q) $64.2^{3x} = 1$

(5) Fig. (viii) shows two right-angled triangles, ABC and BCD.

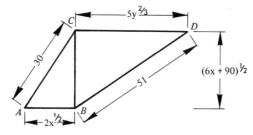

Fig. (viii)

AB, BC, CA, CD and BD have lengths of $2x^{\frac{1}{2}}$, $(6x + 90)^{\frac{1}{2}}$, 30, $5y^{\frac{2}{3}}$ and 51 cm respectively.
Find the values of x and y.

Variation

(1) (a) y varies as x. If $y = 27$ when $x = 18$, calculate the value of y when $x = 28$.

(b) y varies inversely as x. If $y = 40$ when $x = 30$, calculate the value of y when $x = 8$.

(c) y varies as the square root of x. If $y = 8$ when $x = 4$, calculate the value of y when $x = 81$.

(d) *a* varies as the square of *b*. If $a = 28$ when $b = 2$, calculate the value of *a* when $b = 5$.

(e) *a* varies inversely as the square root of *b*. If $a = 0.25$ when $b = 25$, calculate the value of *a* when $b = 16$.

(f) $p \, \alpha \, q^3$. If $p = 28$ when $q = 2$, calculate the value of *q* when $p = 224$.

(g) *y* varies inversely as the square of *x*. If $y = \frac{9}{32}$ when $x = 4$, calculate the values of *x* when $y = \frac{1}{8}$.

(h) *p* varies jointly with *q* and *v*. If $p = 24$ when $q = 2$ and $v = 3$, calculate the value of *p* when $q = 9$ and $r = \frac{1}{2}$.

(i) *E* varies as the square of *v* and as *m*. If $E = 480$ when $m = 60$ and $v = 4$, calculate the value of *E* when $m = 120$ and $v = 2$.

(j) *S* varies as the square of *t* and as the square of sin θ. If $S = 50$ when $t = 2$ and $\theta = 30°$, calculate the value of θ ($0° < \theta < 90°$) when $s = 600$ and $t = 4$.

(2) The time of revolution *T* of a planet around its sun varies as $D^{\frac{3}{2}}$, where *D* is its distance from its sun. Planet *X* takes 365 of its days to revolve about its sun and is 1.5×10^8 km distant from it. Planet *Y* is 10^8 km from the same sun. Calculate (in planet *X* days) the time for planet *Y* to revolve about the sun.

(3) The horizontal distance travelled by a projectile varies as the square of its starting velocity and as the sine of twice the angle of projection. When the starting velocity is 100 m/s and the angle of projection is 15°, the horizontal distance is 500 m. What is the horizontal distance for a starting velocity of 200 m/s and an angle of projection of 30°?

2 Geometry

Triangle—*Specification, Pythagoras*

(1) Show by sketching how two congruent triangles can be obtained from:
    (*a*) A rectangle   (*b*) A square   (*c*) An isosceles triangle
    (*d*) A kite   (*e*) A rhombus

(2) In $\triangle PQR$, $PR = 3$ cm, $QR = 2$ cm and $\angle PRQ = 60°$. Which of the following triangles (not drawn to scale) are congruent to $\triangle PQR$?

(*a*)
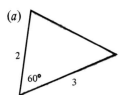

(*b*)

(*c*)
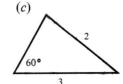

(*d*)
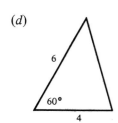

(*e*)
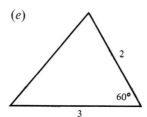

(3) (*a*) State the condition so that a triangle of sides *a* cm, *b* cm and *c* cm may be constructed.
    (*b*) State which of the following triangles can be constructed, and make a construction in those cases.
      (i) $\triangle ABC$: $AB = 7$ cm, $BC = 6$ cm, $CA = 5$ cm
     (ii) $\triangle XYZ$: $XY = 3$ cm, $YZ = 4$ cm, $XZ = 1$ cm
    (iii) $\triangle PQR$: $PQ = 3{\cdot}5$ cm, $PR = 1{\cdot}5$ cm, $QR = 4{\cdot}5$ cm
    (iv) $\triangle LMN$: $LM = 4{\cdot}5$ cm, $MN = 3{\cdot}5$ cm, $LN = 5{\cdot}5$ cm

(4) A plane flies 200 km from town A to town B on a bearing $060°$ and then 300 km on a bearing $150°$ to town C. Using a scale drawing, find the distance in km between A and C. Verify your answer by calculation.

(5) Calculate the length of the third side of each of the following right-angled triangles, not drawn to scale.

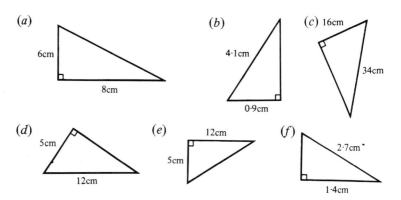

(a) 6cm 8cm

(b) 4·1cm 0·9cm

(c) 16cm 34cm

(d) 5cm 12cm

(e) 12cm 5cm

(f) 2·7cm 1·4cm

(6) A cuboid measures 15 cm by 8 cm by 4 cm. Calculate the lengths of its space diagonals.

(7) A rhombus $ABCD$ has diagonal BD of length 4 cm. If the area of the rhombus is 24 cm², find the lengths of the sides of the rhombus.

(8) A kite $ABCD$ has a diagonal BD of length 12 cm, and its area is 36 cm². If the diagonal BD is divided in the ratio $1:2$ by the diagonal AC, find the length of each side of the kite.

(9) Show that the points A $(1, 2)$, B $(6, 1)$ and C $(3, 4)$ are the vertices of a right-angled triangle.

(10) Find the area of a rectangle $ABCD$, given that AB is of length 12 cm and the diagonal BD is of length 12·5 cm.

(11) State which of the following are right-angled triangles, and name the right-angle.
(a) $\triangle ABC$, with $AB = 3$ cm, $BC = 4$ cm and $AC = 5$ cm.
(b) $\triangle PQR$, with $PQ = 13a$ cm, $PR = 5a$ cm and $RQ = 12a$ cm.
(c) $\triangle LMN$, with $LM = \sqrt{3}$ cm, $LN = \sqrt{4}$ cm and $MN = \sqrt{5}$ cm.
(d) $\triangle HIJ$, with $HI = 5\sqrt{2}$ cm, $IJ = 5$ cm and $JH = 5$ cm.
(e) $\triangle XYZ$, with $XY = 5$ cm, $YZ = 5$ cm and $XZ = 10$ cm.
(f) $\triangle DEF$, with $DF = (a^2 + b^2)$ cm, $EF = (a^2 - b^2)$ cm and $DE = 2ab$ cm.

66

Similarity

(1) State which of the following pairs of triangles are similar.
(Diagrams are not drawn to scale.)

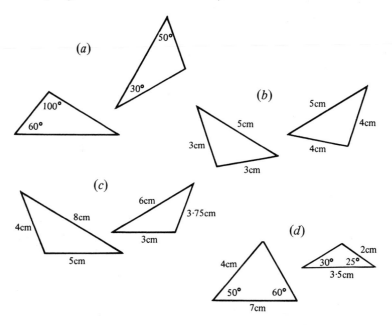

(2) State which of the following pairs of quadrilaterals are similar.
(Diagrams are not drawn to scale.)

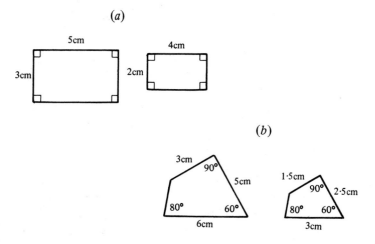

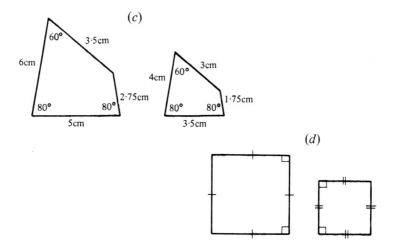

(c)

(d)

(3) In Fig. (i) *ED, DF, FE* are parallel to *BA, AC* and *CB* respectively. If *EF* = 4 cm, *DF* = 3 cm and *BC* = $6\frac{2}{3}$ cm, find the length of *AC*.

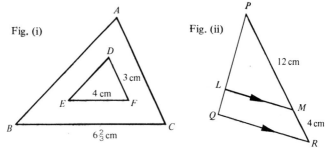

Fig. (i)

Fig. (ii)

(4) In Fig. (ii) *LM* is parallel to *QR*. If *PM* = 12 cm, *MR* = 4 cm and the area of $\triangle PLM$ is 36 cm^2, find the area of $\triangle PQR$.

(5) In Fig. (iii) angle *B* = 90° = angle *AED*,
angle *DBC* = θ, *AE* = 8 units, *EB* = 4 units
and *ED* = 6 units.
(*a*) Name two similar triangles.
(*b*) Calculate the length of *BC*.
(*c*) Calculate the length of *AC*.
(*d*) Name another angle equal to θ.
(*e*) Find the value of tan θ.
(*f*) Calculate the value of the ratio $\dfrac{\triangle AED}{\triangle ABC}$

in its simplest form.

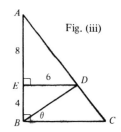

Fig. (iii)

(6) In Fig. (iv) (not drawn to scale) *BE* and *CD* are perpendicular to *AC* and *AB* respectively. *AD* = 9 cm, *BD* = 4 cm and *AE* = 7·8 cm.

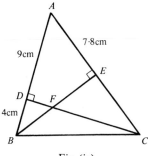

Fig. (iv)

 (*a*) If *S* is a set of similar triangles in the diagram, find *n* (*S*) and list the members of *S*.
 (*b*) Find the lengths of: (i) *BE*   (ii) *DF*   (iii) *BF*
          (iv) *EF*   (v) *FC*   (vi) *EC*

Transformations

Reflection, Bilateral Symmetry

(1) In Fig. (v) *B'*, *D'* are the images of *B* and *D* under reflection in the line *L*.

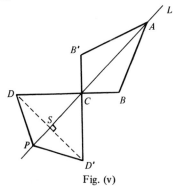

Fig. (v)

 (*a*) Name line segments equal in length to *AB*, *D'C* and *PD*.
 (*b*) Name angles equal to ∠*CAB* and ∠*CPD'*.
 (*c*) *DD'* = 2*. What is *?

(2) State the coordinates of the images P', Q', R', S', T' of the points $P(3, -2)$, $Q(4, 0)$, $R(-1, 3)$, $S(0, -1)$ and $T(-5, 4)$ under reflection in:
(a) The y-axis   (c) The line $x = -3$
(b) The x-axis   (d) The line $y = -1$

(3) State the reflections which map:
(a) $A(2, 0)$ onto $A'(2, -4)$ (b) $B(-3, -2)$ onto $B'(-3, 6)$
(c) $C(5, 1)$ onto $C'(1, 5)$ (d) $D(a, b)$ onto $D'(-a, b)$
(e) $E(-b, a)$ onto $E'(-b, -a)$

(4) Sketch the following figures, showing their axes of bilateral symmetry, if any.
(a) Rectangle   (b) Parallelogram   (c) Rhombus   (d) Kite

(5) Copy the following figures, showing their axes of bilateral symmetry, if any.

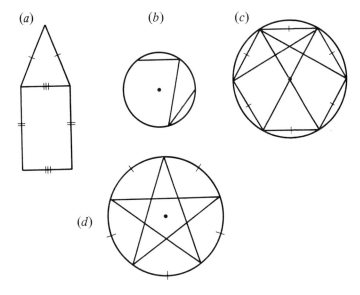

(a)   (b)   (c)

(d)

(6) Plot the points $A(-1, 1)$, $B(-3, -1)$ and $C(1, -1)$ on a Cartesian diagram:
(a) Show the image $\triangle A'B'C'$ of $\triangle ABC$ under reflection in the line $y = -1$. What kind of quadrilateral is formed by $\triangle ABC$ and its image?
(b) Show the image of this quadrilateral under reflection in the line $x = -2$.

Translation

(1) State the coordinates of the images of $A(1, 5), B(-2, 0)$,
$C(-3, -4)$ and $D(5, 6)$ under each of the translations:

(a) $T_1\begin{pmatrix}1\\5\end{pmatrix}$　(b) $T_2\begin{pmatrix}-2\\3\end{pmatrix}$　(c) $T_3\begin{pmatrix}0\\-4\end{pmatrix}$　(d) T_3 followed by T_2

(2) This diagram shows a tiling of congruent triangles.

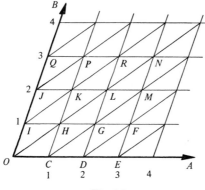

Fig. (vi)

(a) Name the image of $\triangle CHG$ under a translation $\overrightarrow{IP}$.
(b) Name the image of $\triangle HKL$ under a translation $\overrightarrow{KQ}$.
(c) Name the image of $\triangle RPK$ under a translation $\overrightarrow{NG}$.

(d) Name the image of $\triangle IHK$ under a translation $\begin{pmatrix}1\\0\end{pmatrix}$ referred to
oblique axes OA and OB.
(e) Which translation will map $\triangle DEF$ onto $\triangle PQJ$?

(3) Name the translations which map:
(a) $A(2, 3)$ onto $A'(-1, 7)$
(b) $B(a, b)$ onto $B'(a + 3, b)$
(c) $C(2p, 3)$ onto $C'(p, p)$
(d) $D(l + 1, k)$ onto $D'(l, k + 1)$

(4) Plot the points $A(5, 1), B(10, 3)$ and $C(6, 4)$ on a Cartesian
diagram. Mark $\triangle A'B'C'$, the image of $\triangle ABC$ under a translation
$T\begin{pmatrix}-3\\-2\end{pmatrix}$ followed by a reflection in the x-axis.

71

(1) Fig. (vii) shows a regular hexagon
ABCDEF with centre *O*.
(*a*) What rotation about *O* maps:
    (i) *A* onto *B*
    (ii) *A* onto *D*
    (iii) *A* onto *F*

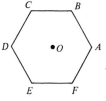

Fig. (vii)

(*b*) Find * so that under a rotation of 120° anti-clockwise about
O:
    (i) $A \rightarrow *$ (iv) $AB \rightarrow *$ (vii) $AD \rightarrow *$
    (ii) $D \rightarrow *$ (v) $FB \rightarrow *$ (viii) $\angle ODE \rightarrow *$
    (iii) $E \rightarrow *$ (vi) $EA \rightarrow *$ (ix) $\angle FAE \rightarrow *$
(*c*) If *P* is the mid-point of *OD*, state the location of *P*′, the image
of *P*, under a rotation of 240° anti-clockwise about *O*.
(*d*) Which point is invariant under each of the above rotations?

(2) For the following state the centres of symmetry (if they exist)
and the angle of rotational symmetry.
(*a*) Equilateral triangle   (*b*) Scalene triangle   (*c*) Rectangle
(*d*) Parallelogram   (*e*) Circle   (*f*) Rhombus

(3) Copy and complete the following table:

Point	Image under rotation of 90° anti-clockwise about the origin	Image under rotation of 180° anti-clockwise about the origin	Image under rotation of 90° clockwise about the point (2, −1)
(3, 0)			
(−4, 0)			
(2, 1)			
(1, −3)			
(1, −1)			
(3, −2)			
(−1, 3)			

(4) Plot the points *A* (2, 6), *B* (2, 2) and *C* (5, 2).
(*a*) Draw *A′B′C′*, the image of triangle *ABC* under an anti-
clockwise rotation of 90° about (i) *B* and (ii) the origin.
(*b*) On a separate Cartesian diagram show *A″B″C″*, the image of
triangle *A′B′C′* under a half-turn about (i) *B* and (ii) the
origin.

(5) A rotation maps *A* (7, 1) onto *A′* (3, 5) and *B* (9, 3) onto *B′* (1, 7).
Find:
(*a*) The centre of rotation.
(*b*) The size of the angle of rotation.

(6) A is $(1, 1)$ and B is $(4, 3)$.
   (a) (i) Find the coordinates of A', the image of A under a half-
      turn about B.
      (ii) Find the coordinates of B', the image of B under a half-
      turn about A.
   (b) (i) $AA' = *BB'$; (ii) $A'B' = *AB$.
      What is $*$ in each case?
   (c) Write down the coordinates of A'' and B'', the images of A'
      and B' under a half-turn about the origin.
   (d) What shape is the quadrilateral $A'B''A''B'$?

Dilatation

(1) With centre of enlargement O and scale factor k, as indicated,
   sketch each of the following and construct their enlargements.

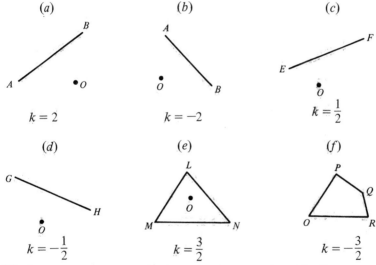

(a)

$k = 2$

(b)

$k = -2$

(c)

$k = \dfrac{1}{2}$

(d)

$k = -\dfrac{1}{2}$

(e)

$k = \dfrac{3}{2}$

(f)

$k = -\dfrac{3}{2}$

(2) With respect to a dilatation, scale factor -1 and centre of
   similitude $P(1, 1)$;
   (a) $\overrightarrow{PR'} = k\,\overrightarrow{PR}$. What is k?
   (b) Find the image of $A(3, 2)$ under this dilatation.
   (c) Find the image of $A(3, 2)$ under a half-turn about P.
   (d) What do you conclude from b and c?
   (e) Write down the image of $B(2, 1)$ under this dilatation.
   (f) Is $\triangle ABC$ congruent to its image $\triangle A'B'C'$?
   (g) Is it true that distances are conserved?

73

(3) A is the point $(2, 1)$. Show on a Cartesian diagram the image of $\triangle ACB$, where B and C are the points $(4, 3)$ and $(2, 4)$ respectively, under a dilatation $[A, 2]$. The images of B and C are D and E respectively.
Show also the image of $\triangle ADE$ under the dilatation $\left[A, -\dfrac{1}{2}\right]$. The images of D and E are F and G respectively.
Find replacements for * in each of the following:
(a) $\overrightarrow{AF} \simeq * \overrightarrow{AB}$ (b) $\overrightarrow{AE} \simeq * \overrightarrow{AC}$
(c) Area of $\triangle AED$ is * times the area of $\triangle ABC$
(d) $FD = * AB$ (e) $\underline{BC} = * \underline{DE}$
(f) $\angle AGF = \angle *$ (g) $\overrightarrow{AC} \simeq * \overrightarrow{GE}$
(h) $\triangle AGF$ is the image of $\triangle AED$ under dilatation $[*, *]$.

(4) State the centre of dilatation in each of the following dilatations:
(a) $A, (3, 1) \rightarrow A'(6, 2)$ $k = 2$
(b) $B(-3, 1) \rightarrow B'(1, -1)$ $k = 3$
(c) $C(-3, 0) \rightarrow C'(-2, 1)$ $k = \dfrac{1}{2}$
(d) $D(1, 1) \rightarrow D'(5, 5)$ $k = -1$
(e) $E(4, 1) \rightarrow E'(7, 4)$ $k = -\dfrac{1}{2}$

(5) State the scale factor in each of the following dilatations:
(a) $A(3, 3) \rightarrow A'(5, 5)$ Centre $(2, 2)$
(b) $B(6, -1) \rightarrow B'(7, 1)$ Centre $(4, -3)$
(c) $C(2, 4) \rightarrow C'(-2, 0)$ Centre $(0, 2)$
(d) $D(2, 2) \rightarrow D'(1, -5)$ Centre $(0, -4)$
(e) $E(2a, a) \rightarrow E'(-2a, a)$ Centre (a, a)

(6) Fig. (viii) shows a tiling of congruent rectangles.

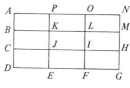

Fig. (viii)

(a) State the image of rectangle $IFGH$ under:
   (i) $[G, 2]$ (ii) $[G, 3]$
(b) State a dilatation which maps:
   (i) Rectangle $ACIO$ on to $IFGH$
   (ii) Rectangle $CDEJ$ on to $OLMN$
   (iii) Rectangle $KJIL$ on to $ILKJ$

(7) Fig. (ix) shows a tiling of congruent triangles:

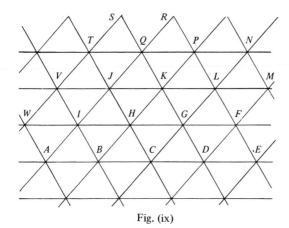

Fig. (ix)

(a) State the image of:
  (i) $\triangle GCD$ under dilatation $[D, 2]$
  (ii) $\triangle JAC$ under dilatation $[J, \frac{1}{2}]$
  (iii) $\triangle PBE$ under dilatation $[E, \frac{1}{3}]$
  (iv) $\triangle JAC$ under dilatation $[H, -1]$
  (v) $\triangle GIQ$ under dilatation $[G, -\frac{1}{2}]$
(b) State in each case the dilatations which map:
  (i) $\triangle AIB$ onto $\triangle ARE$ (ii) $\triangle DFE$ onto $\triangle JHK$
  (iii) Parallelogram $HKLG$ onto parallelogram $AJLC$
(c) $\triangle DFE \rightarrow \triangle JCL$ under the composite transformation $T_1 \circ T_2$, where T_1, T_2 are dilatations. Describe fully T_1 and T_2.

(8) Find in each case the centre of dilatation for the dilatations which map:
(a) $A(0, 3) \rightarrow A'(1, 4)$ and $B(2, 1) \rightarrow B'(4, 1)$
(b) $C(2, 6) \rightarrow C'(6, 2)$ and $D(5, 5) \rightarrow D'(3, 3)$
(c) $E(3, 4) \rightarrow E'(2, 3)$ and $F(3, 3) \rightarrow F'(2, 1)$

Vectors

(1) In Fig. (x):
  $\underline{a} - \underline{b} + \underline{c} - \underline{d} + \underline{e} = \underline{x}$
  State the value of x.

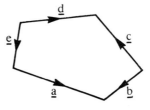

Fig. (x)

(2) (a) $\underline{u} = \begin{pmatrix} 8 \\ 15 \end{pmatrix}$. Calculate its length.

(b) $\underline{u} = \begin{pmatrix} 2 \\ 5 \end{pmatrix}$, $\underline{v} = \begin{pmatrix} -3 \\ -1 \end{pmatrix}$ and $\underline{w} = \begin{pmatrix} 1 \\ 0 \end{pmatrix}$ are vectors. Determine $2\underline{u} - \underline{v} + 3\underline{w}$.

(c) A is $(2, 2)$ and B is $(-2, 1)$. Find the components of $\overrightarrow{AB}$. C is $(6, 3)$. Find the components of $\overrightarrow{AC}$. Hence show that A, B and C are collinear.

(3) In Fig. (xi): $AD:DE:EC = 2:1:1$. Express in terms of p and q the vector represented by:
(a) $\overrightarrow{AE}$ (b) $\overrightarrow{BE}$

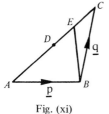

Fig. (xi)

(4) A is $(-2, -3)$ and B is $\left(2, -\dfrac{5}{2}\right)$. Under a half-turn about O, $A \leftrightarrow C$, $B \leftrightarrow D$ and $\overrightarrow{DE} \simeq 2\overrightarrow{DB}$.
(a) Find, in component form:
   (i) $\overrightarrow{DB}$ (ii) $\overrightarrow{DE}$ (iii) $\overrightarrow{OE}$ (iv) $\overrightarrow{OA}$ (v) $\overrightarrow{AE}$
(b) Calculate $|\overrightarrow{AD}|$.

(5) In rectangle $OABC$ in Fig. (xii) $AR:RB = 1:2$, $QR /\!/ OA$ and P is the mid-point of OB. $\overrightarrow{CB}$ represents $\underline{a}$ and $\overrightarrow{OC}$ represents $\underline{b}$. Which vectors do the following displacements represent?
(a) $\overrightarrow{OR}$ (b) $\overrightarrow{PR}$ (c) $\overrightarrow{QR}$ (d) $\overrightarrow{CQ}$

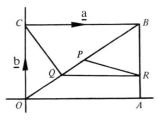

Fig. (xii)

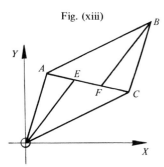

Fig. (xiii)

(6) In Fig. (xiii) $OABC$ is a parallelogram. A and C have position vectors $\underline{a}$ and $\underline{c}$ respectively. $\overrightarrow{AE} \simeq \overrightarrow{EF} \simeq \overrightarrow{FC}$.

(*a*) Write down in terms of **a** and **c** the vectors represented by:

(i) $\overrightarrow{CB}$ (ii) $\overrightarrow{AC}$ (iii) $\overrightarrow{AE}$ (iv) $\overrightarrow{FC}$ (v) $\overrightarrow{OE}$ (vi) $\overrightarrow{FB}$

(*b*) What kind of quadrilateral is *OEBF*?

(7) In Fig. (xiv) *A* and *B* have position vectors **a** and **b** with respect to the origin *O*.

$$AC = CO, \quad AD:DO = 2:1, \quad \overrightarrow{BE} \cong \frac{2}{3} \overrightarrow{BC}$$

Find

(*a*) The position vector of *C*.

(*b*) The position vector of *D*.

(*c*) The vector represented by $\overrightarrow{CB}$.

(*d*) The vector represented by $\overrightarrow{CE}$.

(*e*) The position vector of *E*.

(*f*) The vector represented by $\overrightarrow{DE}$.

(*g*) Using (*f*) write down two statements about $\overrightarrow{DE}$ and $\overrightarrow{OB}$.

Fig. (xiv)

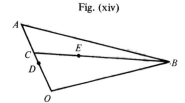

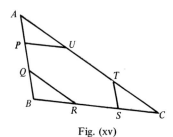

Fig. (xv)

(8) In Fig. (xv) $\overrightarrow{AB}$, $\overrightarrow{BC}$ represent <u>3**a**</u>, 3**b** and 3<u>**c**</u>.

$\overrightarrow{TS}$, $\overrightarrow{PU}$ and $\overrightarrow{RQ}$ represent <u>**a**</u>, <u>**b**</u> and <u>**c**</u>.

(*a*) Why is *PU* ∥ *BC*? Hence name a triangle similar to △*APU*.

(*b*) *AP* = *kAB*. State the value of *k*.

Similarly *BQ* = *lBA*. What is the value of *l*?

Hence *PQ* = *mAB*. What is the value of *m*?

(*c*) $\overrightarrow{PQ}$ + $\overrightarrow{RS}$ + $\overrightarrow{TU}$ represents $\frac{1}{3}$ (**a** + **x** + **y**), which is equal to <u>**z**</u>.

State replacements for **x**, **y** and **z**.

77

Circle, Locus

(1) P is the point $(5, 12)$. Under a half-turn about O, $P \leftrightarrow Q$.
   - (a) State the coordinates of Q.
   - (b) Calculate the length of PQ.
   - (c) Write down the equation of the circle which has PQ as diameter. This circle cuts the x-axis at A and C, and the y-axis at B and D.
   - (d) State the coordinates of A, B, C and D.
   - (e) What size is $\angle APC$?
   - (f) What kind of quadrilateral is $ABCD$?

(2) (a) Show that the point A $(3, 4)$ lies on the circumference of the circle whose equation is $x^2 + y^2 = 25$.
   - (b) State the coordinates of the centre of this circle.
   - (c) Under a positive rotation of $90°$ about the centre of the circle $A \rightarrow B$. State the coordinates of B.
   - (d) Calculate, correct to the second decimal place:
       - (i) The length of the chord AB.
       - (ii) The length of the arc AB.
       - (iii) The area of the triangle OAB.
       - (iv) The area of the sector OAB.

(3) For each of the following diagrams (Fig. (xvi)) find an expression for x, each circle having radius x cm $(x \in R)$.

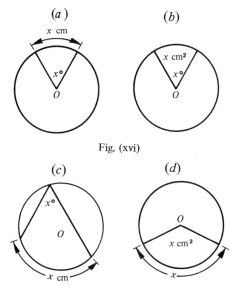

(a)
x cm

(b)
x cm^2
$x°$
O

$x°$
O

Fig. (xvi)

(c)
$x°$
O
x cm

(d)
O
x cm^2
x

(4) Fig. (xvii) shows two concentric circles, centre O. $OA = 33$ cm, $OD = 15$ cm, $\angle AOB = 50°$, and $\angle BOC = 60°$.
Calculate the ratio arc AB: arc DC.

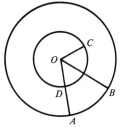

Fig. (xvii)

(5) XY is a chord 8 units long in a circle, centre O and radius 5 units. P is the mid-point of XY. Calculate the length of OP. State the locus of the mid-points of all chords 8 units long in this circle.

(6) $A(-2, 0)$ and $B(2, 0)$ are the end-points of the diameter of a circle. State the equation of the circle of which AB is a diameter.

(7) P is $(2, 6)$ and Q is $(-1, -3)$; A is (x, y) such that $AP = d_1$ and $AQ = d_2$. Write down expressions in simplified form for d_1^2 and d_2^2. Using them, show that the locus of points $A(x, y)$ such that $AP^2 + 2AQ^2 = 108$ is $x^2 + y^2 = 16$. Describe this locus.

(8) Find the equation of the curve which is the locus of a point whose distance from the point $(8, 0)$ is $\frac{8}{17}$ of its distance from the line with equation $x = \frac{289}{8}$.

Sketch the locus, indicating clearly the points where the curve crosses the x- and y-axes.

By drawing two suitable circles, centre O, show that the area bounded by the curve lies between 225π and 289π unit^2.

(9) In Fig. (xviii) AB and AD are the radii of a sphere of diameter 20 cm, and $\angle BAD = 60°$. The sphere rotates about an axis l which is at right angles to AD. C is a point on the axis l such that $\angle ACB = 90°$. State fully and clearly the locus of B.

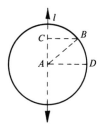

Fig. (xviii)

Miscellaneous Examples

(1) In Fig. (xix)
   $ABCD$ is a rhombus.
   A is $(-4, 3)$.

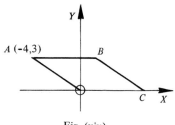

Fig. (xix)

(a) Calculate the length of OA.
(b) Find the coordinates of C.
(c) Calculate the length of AC.
(d) Write in component form the vector represented by $\overrightarrow{BC}$.
(e) Write down the coordinates of B.
(f) What is the value of tan $\angle CBA$?

(2) On a Cartesian diagram mark A (4, 6), B (3, 2) and C (9, 2).
   (a) State the coordinates of M, the mid-point of BC.
   (b) Draw the image of $\triangle ABC$ under a half-turn about M.
      What kind of quadrilateral is formed by $\triangle ABC$ and this image.
   (c) State another transformation which would map $\triangle ABC$ into the image formed in (b).
   (d) Show the image of $\triangle ABC$ under reflection in BC.
      What kind of quadrilateral is formed by $\triangle ABC$ and its image in this case?

(3) In Fig. (xx) A is (2, 3) and B is $(-6, 4)$.

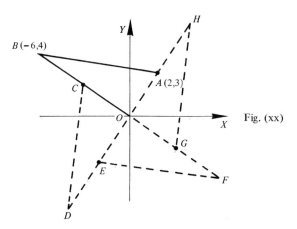

Fig. (xx)

80

(*a*) Use the distance formula to calculate the lengths of:
    (i) *AO*   (ii) *BO*   (iii) *AB*
(*b*) Show that $\angle AOB = 90°$.
(*c*) (i) State the images of *A* and *B* under a counter-clockwise
      rotation of 90° about *O*.
    (ii) State the coordinates of *C* and *D*.
(*d*) The diagram is completed by two further counter-clockwise
    rotations about *O*. State the coordinates of *E, F, G* and *H*.

(4) Plot the points *A* (0, 4) and *B* (6, 2) on a Cartesian diagram.
    A and *B* represent two houses and the *x*-axis a main water-pipe.
    (*a*) Using a reflection, find a point *C* on the *x*-axis such that a
      pipe *ACB* would be of minimum length.
    (*b*) Explain why this point gives a minimum length.
    (*c*) What is this minimum length?

(5) *ABCD* in Fig. (xxi) is a rectangle.
    P is on *BC* and divides *BC* in the ratio 1 : 2.
    PQ is parallel to *DB*, meeting *AB* produced at *Q*.
    PR is parallel to *CA*, meeting *DA* produced at *R*.
    $\overrightarrow{AB}$ and $\overrightarrow{AD}$ represent **u** and **v** respectively.
    (*a*) Show that triangles *DAB* and *PBQ* are similar.

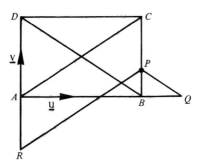

Fig. (xxi)

(*b*) Express $\overrightarrow{PQ}$ in terms of **u** and **v**.
(*c*) Express $\overrightarrow{RQ}$ in terms of **u** and **v**.

81

(6) $A(-5, 10)$, $B(-2, 8)$, $C(-5, 3)$ and $D(-8, 8)$ are the vertices
     of a kite.
     P denotes reflection in the line $x = 4$.
     Q denotes reflection in the line $y = 8$.
     R denotes reflection in the origin.

     (a) Draw a diagram showing $ABCD$, $A_1B_1C_1D_1$, $A_2B_2C_2D_2$ and
         $A_3B_3C_3D_3$, where:
         $A_1B_1C_1D_1$ is the mapping of $ABCD$ under P.
         $A_2B_2C_2D_2$ is the mapping of $ABCD$ under $Q \circ P$.
         $A_3B_3C_3D_3$ is the mapping of $ABCD$ under R.
     (b) Describe fully the single transformation equivalent to $Q \circ P$.
     (c) What single transformation maps $A_2B_2C_2D_2$ into $A_3B_3C_3D_3$?
     (d) Name the coordinates of the centre of rotation which maps
         $A_3B_3C_3D_3$ onto $A_1D_1C_1B_1$.

(7) A is the point $(4, 2)$, B is $(-3, 6)$ and P is the point on OB with
     coordinates $(-1, 2)$.
     (a) Show that $\angle AOB = 90°$.
         Under a half-turn about O, $A \leftrightarrow C$, $B \leftrightarrow D$ and $P \leftrightarrow Q$.
     (b) What kind of quadrilaterals are $ABCD$ and $APCQ$?
     (c) Use the distance formula to show that $BP = PQ = QD$.
     (d) Calculate the ratio,
         area of quadrilateral $APCQ$: area of quadrilateral $ABCD$.

(8) A is the set of isosceles triangles, equilateral triangles, rectangles,
         kites, rhombuses, squares.
     B is the set of right-angled triangles, $30°$, $60°$, $90°$ triangles,
         rectangles, isosceles triangles, scalene triangles, right-angled
         isosceles triangles.
     Illustrate by an arrow diagram the relation "is divided by at least
     one of its axes of bilateral symmetry into two . . ." from set A
     into set B.

(9) $ABCDEFG$ in Fig. (xxii) is a solid
     cube of edge 5 units. By consi-
     dering a suitable net for the cube,
     calculate the shortest distance
     from F to B across the surface.

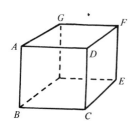

Fig (xxii)

(10) *A* is (4, 6) and *D* is (2, −8). Under a half-turn about the origin,
$A \leftrightarrow C, D \leftrightarrow B$.
   (*a*) What kind of quadrilateral is *ABCD*?
     P, Q, R, S are the mid-points of *OA, OB, OC, OD* respectively.
   (*b*) State the coordinates of *P, Q, R* and *S*.
   (*c*) What kind of quadrilateral is *PQRS*?
   (*d*) Calculate the ratio, quadrilateral *ABCD*: quadrilateral *PQRS*.

(11) $A = \{180°, 120°, 90°, 72°, 60°, 45°\}$
   $B = \{$octagon, equilateral triangle, square, pentagon, parallelo-
       gram, rectangle, square, kite, rhombus $\}$
   Illustrate by an arrow diagram the relation "is an angle of rotational symmetry of . . ." from *A* into *B*.

(12) In Fig. (xxiii) a circle, centre *O*, has a radius of 4 cm, and chord
   AC is of length 5 cm. Calculate the length of chord *BC*.

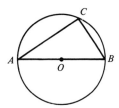

Fig. (xxiii)

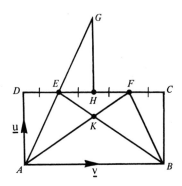

Fig. (xxiv)

(13) In Fig. (xxiv) *ABCD* is a rectangle, and $DE = EH = HF = FC$.
   HG is equal and parallel to *AD*, *EG* is joined, and
   $\overrightarrow{AD}$ and $\overrightarrow{AB}$ represent **u** and **v** respectively.
   (*a*) Express $\overrightarrow{AF}$ and $\overrightarrow{BE}$ in terms of **u** and **v**.
   (*b*) Find (i) the ratio $\dfrac{AK}{KF}$ and (ii) $\overrightarrow{AK}$ in terms of **u** and **v**.
   (*c*) Express (i) $\overrightarrow{AE}$ and (ii) $\overrightarrow{EG}$ in terms of **u** and **v**. Hence show
      that *A, E* and *G* are collinear and *E* is the mid-point of *AG*.

(14) *A* is the point (6, 4) and *B* is (−2, 3).
(*a*) Calculate, using the distance formula or otherwise, the lengths of *OA*, *OB* and *AB*, where *O* is the origin. Leave your answers in surd form.
(*b*) Use your answers to (*a*) to show that ∠*AOB* = 90°.
(*c*) *B'* is the image of *B* under reflection in *OA*. What kind of triangle is △*ABB'*?
(*d*) Calculate the area of △*ABB'*.
(*e*) Using tables, calculate the size of ∠*ABO*.

(15) A cube has a volume of 64 cm³. Find the lengths of its space diagonals.

(16) (*a*) Show on a Cartesian diagram:
(i) △*ABC*, where *A*, *B* and *C* are the points (2, 5), (2, −1) and (0, −1) respectively.
(ii) △*A'B'C'*, the image of △*ABC* under the dilatation [*P*, −1], where *P* is the point (1, 2).
(*b*) What kind of figure is formed by △*ABC* and its image?
(*c*) State an alternative transformation which would map △*ABC* on to △*A'B'C'*.

(17) (*a*) Show on a Cartesian diagram:
(i) The line *AB*, where *A* is the point (3, 2) and *B* the point (−3, 2).
(ii) The line *A'B'*, the image of *AB* under the dilatation [*O*, 2].
(*b*) Find the length of *AB* and hence state the length of *A'B'*.
(*c*) Find the area of △*OAB* and hence state the area of △*OA'B'*.

(18) In Fig. (xxv) *BC* is parallel to *DE*.

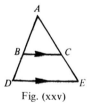

Fig. (xxv)

(*a*) If *AB* = 3 cm and *BD* = 2 cm, state the dilatation which will map:
(i) △*ABC* onto △*ADE*   (ii) △*ADE* onto △*ABC*
(*b*) If *AB* = 4 cm and *BD* = 2 cm, state the dilatation which will map:
(i) △*ABC* onto △*ADE*   (ii) △*ADE* onto △*ABC*

3 Trigonometry

Right-angled Triangles

(1) For each of the diagrams in Fig. (i) write down, in surd form if necessary, the values of $\sin\theta$, $\cos\theta$ and $\tan\theta$.

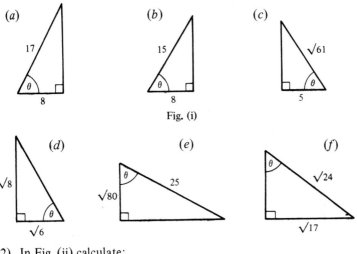

Fig. (i)

(2) In Fig. (ii) calculate:
   (a) $\angle ACB$
   (b) AC
   (c) AB

Fig. (ii)

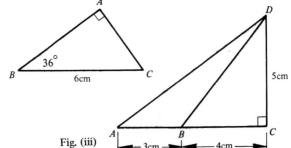

Fig. (iii)

(3) In Fig. (iii) Q is any point on AB between A and B. What is the range of possible values of $\angle DQC$?

(4) *ABCD* is a rhombus such that *CD* = 3 cm and diagonal
AC = 4 cm. Calculate the sizes of the angles of the rhombus.

Sine Rule, Cosine Rule, Area

(1) In △*ABC*, ∠*ABC* = 43°,
∠*BAC* = 102° and *AB* = 13·8 cm.
Calculate the lengths of *AC* and *BC*.

(2) In Fig. (iv), not drawn to scale, calculate the
size of angle *PQR*.

(3) In △*XYZ*, ∠*XYZ* = 40°,
∠*XZY* = 36° and *YZ* = 5 cm.
Calculate the length of *XY*.
P is a point on *YZ* such that
YP : *PZ* = 3 : 2.
Find the length of *XP*.

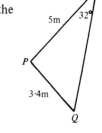

Fig. (iv)

(4) In △*EFG*, *EF* = 12 cm,
EG = 18 cm and ∠*EFG* = 105°.
Calculate the size of ∠*GEF*.

(5) A quadrilateral *ABCD* is inscribed in a circle. *AB* = 8 cm,
AD = 4 cm and ∠*ACB* = 60°. Calculate the length of *BD*.

(6) In △*ABC*, *BC* = 6 cm, *AB* = 8 cm and *AC* = 4 cm. Calculate the
size of angle *ACB*. *BC* is produced to a point *D* such that angle
CAD is 42°. Find the length of *CD*.

(7) In the diagrams in Fig. (v):

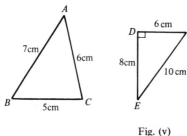
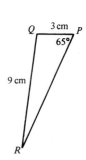

Fig. (v)

What is the area of:
(*a*) △*ABC*, (*b*) △*DEF* and (*c*) △*PQR*?

(8) In △*KLM*, *KL* = 10 cm and *LM* = 8 cm. If the area of this
triangle is 32 cm², calculate the possible sizes of ∠*KLM*.

(9) In Fig. (vi) ∠*ACB* = 90°, ∠*ABC* = 57°, *BC* = 25 cm, *CD* = 56·2 cm
and ∠*CAD* = 110°.

86

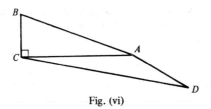

Fig. (vi)

Calculate:
(a) The length of *CA*. (b) The size of ∠*ACD*.
(c) The area of triangle *ACD*.

Three-Dimensional Situations

(1) In Fig. (vii) below *ABCDEFGH* is a cuboid. *AB* = 4 cm,
BC = 3 cm, and *CG* = 12 cm.
Calculate:
(a) The length of *AC*. (b) The length of *BG*.
(c) The length of *AG*. (Leave in surd form.)
(d) The area of △*ABG*. (e) The size of the angle between
(Leave in surd form.) *AG* and the plane *BCGF*.

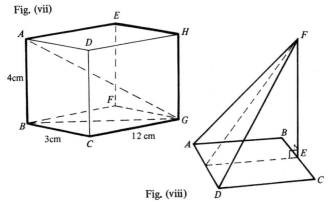

Fig. (vii)

Fig. (viii)

(2) Fig. (viii) shows a rectangular plane *ABCD* in which *AD* = 12 cm
and *DC* = 8 cm. *E* is the mid-point of *BC*, and *EF* is perpendicular
to the plane *ABCD* and has length 24 cm.
Calculate:
(a) The length of *DE*. (b) The length of *DF*.
(c) The angle between *DF* and the plane *ABCD*.
(d) The angle between the planes *ABCD* and *ADF*.

87

(3) Fig. (ix) below shows a vertical pole, *TP*, at a corner *P* of a
horizontal square, *PQRS. TP* and *PQ* have lengths of 5 m and 8 m
respectively. Find:

(a) The length of *TR*.

(b) The angle between *TR* and the plane *PQRS*.

(c) The angle between the planes *TQS* and *PQRS*.

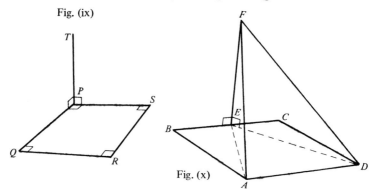

Fig. (ix)

Fig. (x)

(4) Fig. (x) shows a rectangular plane *ABCD*, in which *AB* = 12 cm
and *BC* = 14 cm. At *E*, a point on *BC*, 5 cm from *B*, *EF* is
drawn perpendicular to the plane *ABCD* and of length 20 cm.

(a) Calculate the lengths of:

    (i) *DE*   (ii) *AE*   (iii) *DF*

    (iv) *AF* (Leave answer in surd form.)

(b) Calculate the size of ∠*ADF*.

(5) Fig. (xi) shows a right pyramid *BPQRS* with square base of edge
4 cm and vertical height 5 cm. Find the angle between:

(a) The plane *BRS* and the base.   (b) *BR* and the base.

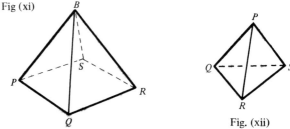

Fig (xi)

Fig. (xii)

(6) Fig. (xii) shows a regular tetrahedron *PQRS* of edge 10 cm. Find
the length of an altitude of a face, and hence the angle between:

(a) The planes *PQS* and *PRS*.   (b) *PR* and the plane *QRS*.

(7) In Fig. (xiii) *PQ* represents a tower 45 m high. From the positions
R and *S* on the ground the angles of elevation to *Q*, the top of the
tower, are $16°$ and $18°$ respectively. If $\angle RPS = 38°$, calculate to
the length of *RS* to the nearest metre.

Fig. (xiii)

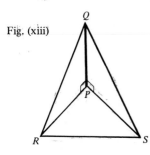

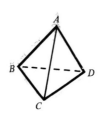

Fig. (xiv)

(8) *ABCD* in Fig. (xiv) is a regular tetrahedron of edge 6 cm.
  (*a*) By considering *M*, the mid-point of *CD*, calculate *AM*.
     Hence write down the value of *BM*.
  (*b*) Hence calculate the angle between the planes *ADC* and *BDC*.

(9) For the slant pyramid *EABCD* in Fig. (xv) with square base
  $AB = BC = CD = DA = 4; EC = ED = 2\sqrt{2};$ and $EA = EB = 2\sqrt{6}.$
  Calculate the angle between:
  (*a*) *EA* and *EC*
  (*b*) *ED* and *EC*
  (*c*) The planes *EDC* and *EAC*

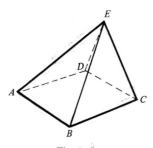

Fig. (xv)

Bearings

(1) (*a*) A ship sails due North from *A* 20 km to a point *B* and then
     sails on a bearing $130°$ for 15 km to a point *C*. Calculate the
     length of *AC*.
  (*b*) Another ship sails from *B* on a bearing $050°$ to a point *D*,
     due North of *C*. Calculate:
     (i) $\angle DBC$ (ii) *BD*   (iii) *DC*

(2) A ship is anchored at a position A. A lighthouse C bears $44°$ East of North from A. The ship sails due East for 21 km to a position B, from which the lighthouse C has a bearing $27°$ West of North from B. Calculate:
   (a) The distance from C to B.
   (b) How far the lighthouse is from the ship at its nearest point.

(3) A ship leaves a port A and sails 24 km to a port B, $45°$ East of North from A. It then changes course to $35°$ West of North and sails until it reaches a point C, which is directly North of A. Calculate the distances of C from A and B.

(4) A is a position 50 km due South of a position B. A ship sails from a port S, which is due West of B, and in this position the bearing of S from A is $305°$. Three hours later the ship is in a position C, due West of A, and in this new position the bearing of C from B is $218°$. Calculate:
   (a) The distances AC and BS, and find the bearing of C from S.
   (b) The speed of the ship.

(5) A ship steams due West at 15 km per hour. At 10.00 hours the ship is 10 km from a lighthouse and at 11.00 hours it is 12 km from the lighthouse. Find the bearing of the lighthouse from the ship
   (a) at 10.00 hours and (b) at 11.00 hours.

(6) From a plane flying on a course $345°$ at a speed of 330 km/h, a pilot observed a landmark on a bearing $280°$. Ten minutes later the bearing of the same landmark was $230°$. How far were the initial and final positions of the plane from the landmark?

Miscellaneous Examples

(1) In Fig. (xvi), which represents a triangular field PQR:
   $\frac{PS}{SQ} = \frac{1}{3}$ and $\frac{PT}{TR} = \frac{1}{2}$.
   $SQ = 42$ m, $TR = 36$ m and $QR = 24$ m.
   Calculate:
   (a) The lengths of PQ and PR.
   (b) The size of angle QPR.
   (c) The area of PST to the nearest m^2.
   (d) The area of $STRQ$ to the nearest m^2.

Fig. (xvi)

90

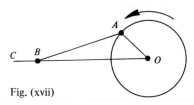

Fig. (xvii)

(2) Fig. (xvii) shows a wheel which revolves about O. AB is a rod attached to the wheel at A and to a fixed rod OC in such a way that, as the wheel revolves, the end B moves backwards and forwards along the rod OC. $OA = 5$ cm and $AB = 12$ cm.

(a) What is the least distance between O and B?

(b) What is the greatest distance that B is from O?

(c) What is the distance OB when $\angle OAB = 90°$?

(d) What is the distance OB when $\angle OAB = 120°$?

(e) What size is $\angle OAB$ when $OB = 11$ cm?

(f) What is the maximum area of the triangle formed by OA, OB and AB?

(3) In Fig. (xviii) the sphere has two circular cross-sections shown. One is through the centre of the sphere O and has radius 20 cm. The other touches the resulting great circle in one point P as shown. This smaller cross-section has radius 10 cm. What is the size of the angle between these circular cross-sections?

Fig. (xviii)

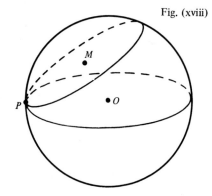

Fig. (xix)

(4) In Fig. (xix) B and D are two points on the North Sea coast. B is due North of D. A pipeline has to be laid along the sea bed from an oilwell at A, either directly to B, or to D along the path ACD.

A lies 100 km from B on a bearing 120°. C lies 130 km from B on a bearing 160°, and is due East of D. By how much does the path ACD exceed the path AB?

91

(5) (a) If $2 \cos x + 1 = 0$, $0° < x < 180°$, what is the value of x?
(b) If $y = 30°$, what is the value of $\sin y \cos^2 x + \sin y \sin^2 x$?

(6) In Fig. (xx) triangles ABD and EBC are right angled at B and with $\angle ADB = 60°$.
$AB = 10$ cm; $AB = BC = x$ cm; $BD = DE = y$ cm;
and $CE = z$ cm.
Calculate:
(a) x, y, E, C and z.
(b) The areas of ABD and EBC.

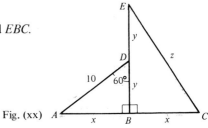

Fig. (xx)

(7) Fig. (xxi) illustrates the positions of two aircraft A and B relative to an airport O.
$OC = 18$ km; $OD = 24$ km; $\angle AOD = 5°$; and $\angle BOC = 10°$.
C and D are positions due East and North respectively of the airport. Calculate:
(a) The difference in the heights of the aircraft.
(b) Their distance AB apart from each other.

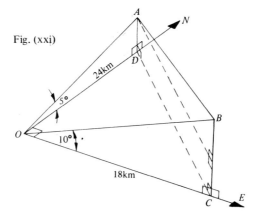

Fig. (xxi)

Appendix

Polar Coordinates

Appendix
Polar Coordinates

Fig. (i) shows a point $P(x, y)$, where the magnitude of $\overrightarrow{OP}$ is r units and angle $XOP = \theta°$. PM is perpendicular to the x-axis.

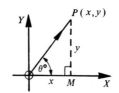

Fig. (i)

Note I x and y uniquely specify the position of P.
The ordered pair (x, y) are called the **Cartesian coordinates** of P for $x, y \in R$.

II $\theta°$ is the magnitude of a positive or anti-clockwise rotation about O, the origin or pole, and r is the magnitude of the displacement $\overrightarrow{OP}$.

III The ordered pair $(r, \theta°)$ are called the **Polar coordinates** of P with respect to the **Pole O** and the initial line OX.

IV The polar coordinates of a point defined as above are NOT unique as the direction of $\overrightarrow{OP}$ can be described as $\theta° + n360°$, where $n \in W$.

V Although it is possible to consider negative values of r, in this section, however, when $(r, \theta°)$ are written as the polar coordinates of a point, it is assumed that:

    (a) $r \geqslant 0$ (b) $0 \leqslant \theta < 360$

Example. In Fig. (ii) $A, B, C, \ldots$ represent the hour positions on a circular clock face of radius 4 units.

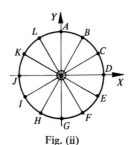

The polar coordinates of D, B and L are $(4, 0°), (4, 60°)$ and $(4, 120°)$ respectively.

Fig. (ii)

Exercise Set

(1) From Fig. (ii) above write down the polar coordinates of C, A, K, J, I, H, G, F and E.

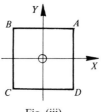

Fig. (iii)

(2) $ABCD$ in Fig. (iii) is a square of sides 2 units and centre O.
(a) Calculate the value of OA, leaving the answer in surd form.
(b) Write down the polar coordinates of A, B, C and D.

(3) The circle $x^2 + y^2 = 16$ cuts the line $y = x$ at points P and Q.
(a) Draw a diagram illustrating this situation.
(b) Write down the polar coordinates of P and Q.

(4) The circle $x^2 + y^2 = 9$ cuts the line $y = -x$ at the points R and S.
(a) Draw a diagram illustrating this situation.
(b) Write down the polar coordinates of R and S,

(5) Fig. (iv) shows the half-plane $y \geqslant 0$ and the lines $y = x$ and $y = -x$.
(a) In which region does the point with polar coordinates $(3, 150°)$ lie?
(i) I   (ii) II   (iii) III   (iv) IV   (v) None of these
(b) If the point with polar coordinates $(r, \theta°)$ lies within region III, which of the following is true?
(i) $0 \leqslant \theta \leqslant 45$ (ii) $45 \leqslant \theta \leqslant 90$ (iii) $90 \leqslant \theta \leqslant 135$
(iv) $135 \leqslant \theta \leqslant 180$ (v) None of these

Fig. (iv)

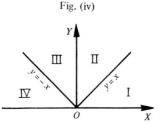

Fig. (v)

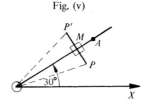

(6) In Fig. (v) P is the point $(4, 20°)$ in polar coordinates with respect to the initial line OX; $\angle XOA = 30°$; P' is the image of P under reflection in OA.

(a) Write down the length of (i) OP and (ii) OP'.

(b) Write down the size of $\angle XOP$.

(c) Calculate the size of $\angle POM$ and hence write down the size of $\angle MOP'$.

(d) What is the size of $\angle XOP'$?

(e) Write down the polar coordinates of P'.

(7) Repeat question (6) for each of the following cases and hence complete the table.

	Angle XOA	Polar coordinates of P	Polar coordinates of image P'
(a)	30°	$(3, 40°)$	
(b)	60°	$(6, 20°)$	
(c)	100°	$(4, 35°)$	
(d)	150°	$(3, 160°)$	
(e)	$a°(0 \leqslant a \leqslant 90)$	$(r, \theta°)\ (\theta < a)$	
(f)	$a°(90 \leqslant a \leqslant 180)$	$(r, \theta°)\ (\theta < a)$	
(g)	$a°(90 \leqslant a \leqslant 180)$	$(r, \theta°)\ (a \leqslant \theta \leqslant 180)$	

(8) In Fig. (vi) P is the point $(r, a°)$ in polar coordinates with respect to origin O and initial line OX; $\angle XOA = 40°$ and $\angle XOB = 85°$.

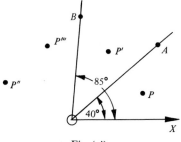

Fig. (vi)

(a) If P' is the image of P under reflection in OA, show that $\angle XOP' = (80 - a)°$.

(b) If P'' is the image of P under reflection in OB, find the size of $\angle XOP''$.

(c) Find the polar coordinates of P''', the image of P' under reflection in OB.

97

(9) Copy and complete the following table:

	Polar coordinates of P	Dilatation	Polar coordinates of image P'
(a)	$(1, 30°)$	$[O, 2]$	
(b)	$(3, 60°)$	$[O, \frac{1}{3}]$	
(c)	$(2, 150°)$	$[O, -1]$	
(d)	$(3, 45°)$	$[O, -2]$	
(e)	$(r, \theta°)$	$[O, k]\ k > 0$	
(f)	$(r, \theta°)\ (0 \leqslant \theta \leqslant 90)$	$[O, k]\ k < 0$	
(g)	$(r, \theta°)\ (90 \leqslant \theta \leqslant 180)$	$[O, k]\ k < 0$	

(10) Under what transformations is P' the image of P in each of the following:

	Polar coordinates of P	Polar coordinates of image P'
(a)	$(3, 30°)$	$(3, 70°)$
(b)	$(3, 30°)$	$(6, 30°)$
(c)	$(3, 30°)$	$(6, 210°)$
(d)	$(2, 90°)$	$(1, 270°)$

Answers

Part I Objective Test Questions

1 MULTIPLE CHOICE ITEMS

Algebra

(1) C.	(2) A.	(3) E.	(4) E.	(5) A.
(6) E.	(7) A.	(8) B.	(9) D.	(10) A.
(11) B.	(12) A.	(13) E.	(14) C.	(15) B.
(16) C.	(17) E.	(18) D.	(19) B.	(20) B.
(21) A.	(22) B.	(23) E.	(24) A.	(25) B.
(26) D.	(27) D.	(28) E.	(29) E.	(30) E.
(31) D.	(32) C.	(33) C.	(34) B.	(35) B.
(36) C.	(37) A.	(38) C.	(39) D.	(40) D.
(41) C.	(42) C.	(43) D.	(44) A.	(45) D.

Geometry

(1) A.	(2) C.	(3) D.	(4) D.	(5) A.
(6) A.	(7) B.	(8) C.	(9) D.	(10) C.
(11) D.	(12) E.	(13) D.	(14) C.	(15) D.
(16) E.	(17) C.	(18) B.	(19) A.	(20) B.
(21) E.	(22) C.	(23) D.	(24) E.	(25) E.
(26) B.	(27) B.	(28) A.	(29) E.	(30) A.

Trigonometry

(1) B.	(2) E.	(3) B.	(4) D.	(5) E.
(6) C.	(7) C.	(8) A.	(9) B.	(10) E.
(11) A.	(12) E.	(13) A.	(14) C.	(15) C.
(16) D.	(17) B.	(18) E.	(19) D.	(20) A.
(21) E.	(22) A.	(23) B.		

2 SITUATION ITEMS

(1)	(i) E.	(ii) A.	(iii) D.	(iv) B.	(v) C.
(2)	(i) B.	(ii) A.	(iii) E.	(iv) C.	
(3)	(i) E.	(ii) A.	(iii) B.		
(4)	(i) C.	(ii) B.	(iii) E.	(iv) A.	(v) D.

3 MULTIPLE COMPLETION ITEMS

Format I

(1) E.	(2) C.	(3) B.	(4) E.	(5) B.
(6) E.	(7) D.	(8) A.	(9) A.	(10) D.
(11) A.	(12) B.	(13) B.	(14) C.	(15) E.
(16) D.	(17) A.	(18) C.	(19) B.	(20) D.
(21) A.	(22) C.	(23) E.	(24) C.	

Format II

(1) D.	(2) A.	(3) A.	(4) A.	(5) E.
(6) E.	(7) C.	(8) B.	(9) D.	(10) E.

Part 2 General Revision Questions

1 ALGEBRA

Sets

(1)   (a) {8, 16, 24, 32, 40, 48}   (b) {31, 37}
      (c) {January, March, May, July, August, October, December}
      (d) {3, 5, 7, 9, 11}

(2)   (b), (f) and (g)

(3)   (a) (i) {6}   (ii) {4, 10}   (iii) {3, 4, 6, 9, 10, 12, 15, 16, 18, 22, 28}
      (iv) {2, 4, 6, 8, 10, 16, 22, 28}
      (b) (i) 5   (ii) 11   (iii) 2

(4)   (a) {p, q, r, s}   (b) {1, 2, 3, 4, 5, 6}   (c) {△, □, ○}
      (d) {×, +, −, ÷}; {×, +, −, ÷}

(5)

```
 E
        ┌──── A ─────   B ────┐
        │  •18     ┌────┐     │
        │        •12│  • 6    │
        │ •21       │         │
        │        •15│  • 9    │
        │    •24  └────┘      │
   •27  └─────────────────────┘
                          • 3
```

(6)   (a) (i) {3, 6, 9, 27}   (ii) {3, 18, 21, 24, 27}   (iii) {18, 21, 24}
      (iv) {3, 6, 9, 12, 15, 27}   (v) {3, 6, 9, 27}
      (vi) {12, 15, 18, 21, 24}
      (b) (i) 5   (ii) 15; 20

(7)   (a) (i) B' (ii) $A \cap B'$ (iii) $(A \cap B)'$
      (b) (i) (x) and (y) and (z)   (ii) None   (iii) (x) and (y)

(8)   (i) $A \cup B \cup C$ (ii) $(A \cap B \cap C)'$ (iii) $(A \cup B \cup C)'$
      (iv) $(B \cup C)'$ (v) C' (vi) $(A \cap C) \cup (B \cap C)$ or $(A \cup B) \cap C$

(9) $E \cup A = E; E \cap A = A; A \cup A = A; A \cap A = A; A \cup A' = E; A \cap A' = \phi;$
      $E \cup A' = E; E \cap A' = A'$

(10)

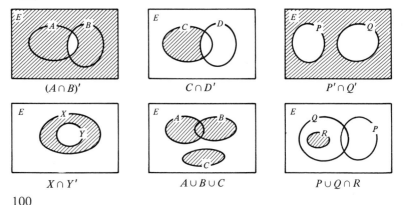

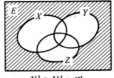

$X' \cap Y' \cap Z'$

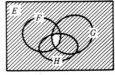

$F' \cup G' \cup H'$

(11)   (a) 12   (b) 10   (c) 10   (d) 6   (e) 4   (f) 6   (g) 2   (h) 3
      (i) 6

(12)   (a) People who like walking only.
      (b) People who prefer buses and who have a bicycle and dislike walking.
      (c) People who prefer buses, like walking but who have not a bicycle.

(13)   (a) 2   (b) Q   (c) $* = j$ (d) P or E   (e) R   (f) R

Laws of Algebra—Commutative and Associative Laws

(1)   (a) 2   (b) −7   (c) −12   (d) −59   (e) 2   (f) $1\frac{1}{6}$ (g) $-3\frac{1}{6}$

(2)   (a) $\frac{1}{3}$ (b) $-2\frac{2}{3}$ (c) $-3\frac{2}{3}$ (d) $3\frac{1}{3}$; No

(3)

*	−2	2
−2	−2	2
2	2	−2

(4)   (a), (b), (c) and (e) are true.

(5)   (a) (i) 25   (ii) 25   (iii) −10   (iv) −10   (v) 185   (vi) 165   (vii) 90a
        (viii) 130a
      (b) (i) Yes (ii) No   (c) (i) Yes (ii) No

Distributive Law

(1)   (a) $3a + 3bc$ (b) $4 - 4b$ (c) $xw - xyz$ (d) $-p^2 - pq$
      (e) $xw + yw$ (f) $-4x^2y + 3xy^3$ (g) $5px + 5qrx$ (h) $-7x^2 + 7yx$
      (i) $3ab^2 - 3b^3$

(2)   (a) $3(a + b)$ (b) $a(a + b)$ (c) $-y(x + z)$ (d) $3xy(x + 2y)$
      (e) $x(x - 2y - 3x^2)$ (f) $a + bc + def$ (g) $-x^2(1 - y)$
      (h) $x^2(1 + y)$ (i) $xy^2(x - 1)$

(3)   (a) $6(4x + 3y)$ (b) $-4(x + y)$ (c) $x^2(9 + x^2)$
      (d) $(x + y)(x + y + 1)$ (e) 0   (f) $a(x - y)(x + y)$
      (g) $2(2x + 1)(2x - 1)$ (h) $(2x + 3y)(2x - 3y)$ (i) $(3x - y)(3x + y)$
      (j) $(x + y)(x + y + 1)(x + y - 1)$

(4)   (a) $3(a + 2b)$ (b) $4(3p - 4q)$ (c) $15(1 - 3ab)$ (d) $a(b - c)$
      (e) $2x(2y - 3z)$ (f) $3a(a - 3b)$ (g) $5q(2p - r)$ (h) $3x(x - 2y)$
      (i) $xy(x - y)$ (j) $7(a + 2b - 3c)$ (k) $ab(a + b - c)$ (l) $3(2 - ab + 4b)$
      (m) $a(a - 1)$ (n) $a^2 + 4b$ (o) $2xy(1 - 2x)$ (p) $36(2a^2 - 3c)$
      (q) $a(6ab - 5d)$ (r) $(x + y)(2x + 2y - 3)$

(5)   (a) $x^2 + 5x + 6$ (b) $x^2 + 12x + 35$ (c) $a^2 - 9a + 20$
      (d) $m^2 - 8m + 7$ (e) $a^2 + 2a - 3$ (f) $a^2 - 2a - 15$
      (g) $p^2 - p - 110$ (h) $p^2 - 49$ (i) $m^2 - 9$
      (j) $6x^2 + 13x + 6$ (k) $6x^2 - 13x + 6$ (l) $20x^2 + 9x - 20$
      (m) $12x^2 - 25x - 7$ (n) $1 - x - 12x^2$ (o) $2x^2 - xy - 6y^2$
      (p) $6a^2 - 5ab - 6b^2$ (q) $a^2 + 2ab + 2ac + 4bc$
      (r) $9ac + 3ab + 6bc + 2b^2$ (s) $7a^4 - 4a^2b^2 - 3b^4$
      (t) $9x^4 - 4y^2$ (u) $25a^2 - 4b^2$

(6) (a) $x^2 + 2xy + y^2$ (b) $9x^2 + 2xy + 4y^2$ (c) $25a^2 + 10ab + b^2$
     (d) $p^2 + 6pq + 9q^2$ (e) $x^2 - 2xy + y^2$ (f) $4x^2 - 12xy + 9y^2$
     (g) $16x^2 - 24x + 9$ (h) $25 - 10x + x^2$ (i) $4 - 12x + 9x^2$
     (j) $1 - 16x + 64x^2$ (k) $25a^2b^2 - 30ab + 9$ (l) $9a^4 + 12a^2b + 4b^2$
     (m) $36 + 2xy + x^2y^2$ (n) $\pi^2 - 4\pi + 4$ (o) $a^2c^2 - 6ab^2c + 9b^4$

(7) (a) 961   (b) 841   (c) 9 801   (d) 10 201
     (e) 1·21   (f) 3·61   (g) 1·0201   (h) 0·980 1

(8) (a) $11x + 5y$ (b) $2x^2 + 13x + 29$ (c) $-a + 13$
     (d) $6x^2 + 20x - 1$ (e) $-43ab + 9b^2$ (f) $25x^2 - 36x + 13$
     (g) $-30x^2 + 12x - 19$ (h) $25x^2 - 36xy + 6y^2$ (i) $-171 + 408p - 241p^2$
     (j) $-4x^2 + 25$ (k) $15x^2 + 32x + 4$ (l) $8 - 4x^2 - 7xy + 15y^2$
     (m) $-2x^2 + 12xy - 4y^2$ (n) $-4x + 7$

(9) (a) $(a + 2)(a + 1)$ (b) $(x - 2)(x - 1)$ (c) $(p - 4)(p + 1)$
     (d) $(m + 4)(m - 1)$ (e)   (f) $(3a + 1)(a - 1)$
     (g) $(5m - 4)(m + 2)$ (h) $(3x + y)(x + 3y)$ (i) $(4x - y)(x - 4y)$
     (j) $(3a - 2)(2a + 3)$ (k) $(6a - 1)(a - 6)$ (l) $(4x + 3y)(3x - 4y)$
     (m) $(6x - 7y)(2x + 5x)$ (n) $(12x + y)(x - 12y)$
     (o) $(6x - y)(2x + 15y)$ (p) $4x(3x + 2y)(2x - 5y)$
     (q) $a^2b(2a + 3b)(a + 2b)$ (r) $(a + b - 2)(a + b - 1)$

(10) (a) $(a + 5)(a + 3)$ (b) $(x + 7)(x + 2)$ (c) $(a + b)(a - 3)$
     (d) $(m + 6)(m - 5)$ (e) $(a - 5)(a - 4)$ (f) $(x - 2)(x - 1)$
     (g) $(p + 2q)(p + q)$ (h)   (i) $(x - 6y)(x - 2y)$
     (j) $(1 - 2y)(1 - 6y)$ (k) $(m - 4n)(m - n)$ (l)
     (m) $(4a + 3)(3a + 4)$ (n) $(5x - 3)(3x + 4)$ (o) $(6x - 5y)(x + 2y)$
     (p) $(5c - 4d)(3c + 5d)$ (q) $(5a^2 - 3)(2a^2 + 1)$ (r) $(5 + 3a)(4 - a)$
     (s) $(6 + p)(1 + 6p)$ (t) $(6a + b)(2a - 3b)$ (u)
     (v) $2(4a - 3)(2a - 3)$ (w) $3(3 - 4m)(2 - 3m)$
     (x) $(2x^2 - 3y^2)(x^2 + y^2)$

(11) (a) $(x + 2y)(x + 2y)$ (b)   (c) $(4x + 1)(3x + 2)$
     (d) $(5y + 2)(3y + 4)$ (e) $(3x - 2y)(2x - 3y)$ (f) $(3a - 7b)(4a + 3b)$
     (g) $(6x + 5)(4x - 3)$ (h) $6(2x + 1)(x - 2)$ (i) $(4x + 3y)(3x + 4y)$
     (j) $a^2(6 + 7x)(5 + 2x)$ (k) $x(x - 1)$ (l) $(y - 1)(y + 4)$

(12) (a) $-3; 11; -7; 21$ (b) $2; -8; 7; -25$ (c) $5; -3; 39; 11$

Difference of Squares

(1) (a) $7(1 - q)(1 + q)(1 + q^2)$ (b) $3b(a - 3b)(a + 3b)$
     (c) $\left(\dfrac{a}{2} - 2b\right)\left(\dfrac{a}{2} + 2b\right)$ (d) $(a - b + c)(a + b - c)$

(2) (a) $(x - y)(x + y)$ (b) $(1 - 2b)(1 + 2b)$ (c) $(3a - 2b)(3a + 2b)$
     (d) $(5a - 7b)(5a + 7b)$ (e) $(11 - 6x)(11 + 6x)$
     (f) $(10a - 9c)(10a + 9c)$ (g) $(x - y)(x + y)(x^2 + y^2)$
     (h) $x^2y^2(x - y)(x + y)$ (i) $x^2y^2(x - y)(x + y)(x^2 + y^2)$
     (j) $x^4y^4(4x^4 + 9y^4)$ (k) $(2a - 3b)(2a + 3b)(4a^2 + 9b^2)$
     (l) $(5 - x)(5 + x)(25 + x^2)$ (m) $9(9x^2 + 4y^2)$
     (n) $a(a - b)(a + b)$ (o) $5(2x - y)(2x + y)$
     (p) $x^2y^2(x - y)$ (q) $7(3a - 2b)(3a + 2b)$
     (r) $2x(2x - 3y^2)(2x + 3y^2)$

(3) (a) 12   (b) 16   (c) 9   (d) 44·2   (e) 1225   (f) 3·38

Miscellaneous

(1) $(3x - 10)(x + 1)$ (2) $xy(y - x)(y + x)$ (3) $2k(1 + 9k)$

(4) $(5 - 2a)(3 - a)$ (5) $10(y - 3)(y + 1)$ (6) No factors

(7) $(2x + 2y - 1)(2x + 2y + 1)$ (8) $(a + b)(a + b + 3)$

(9) $(2z - 7)(3z + 8)$ (10) $4 \cos A (2 \cos A - 1)$

(11) $(\tan A - 2)(3 \tan A + 2)$

(12) $5(\sin A - 2 \cos A)(\sin A + 2 \cos A)$ (13) $(3 - c)(1 + 4c)$

(14) $8pq$ (15) $(p - 2)(p - 1)(p + 1)(p + 2)$

(16) (a) $(l + 2m)(l - 2m)$

       (b) $(l = 5)(m = \tfrac{1}{2})$

Algebraic Fractions

(1)   (a) 5 (b) $\dfrac{5y}{2}$ (c) $\dfrac{ac}{2}$ (d) $\dfrac{4p^2}{3q^2}$ (e) $\dfrac{3}{xyz}$

(2)   (a) ax (b) $4y^2$ (c) $16qx$ (d) $20xy^2$

(3)   (a) (i) $\dfrac{3p}{4}$ (ii) $\dfrac{x}{2}$ (iii) $\dfrac{y - 10}{5}$ (iv) $\dfrac{5a}{8}$

       (b) (i) $\dfrac{2x + 5}{5x}$ (ii) $\dfrac{5}{2a}$ (iii) $\dfrac{x + y}{2xy}$ (iv) $\dfrac{c^2 + d^2 + cd}{cd}$

         (v) $\dfrac{6x^2 + 1}{4xy}$ (vi) $\dfrac{5 + 2a^2b}{2a}$ (vii) $\dfrac{2p^2 - 4p + 1}{p}$

         (viii) $\dfrac{a^2 + b^2 + 2ab}{ab}$

(4)   (a) $\dfrac{3x + 2}{x^2}$ (b) $\dfrac{6p + 5q}{4q^2}$ (c) $\dfrac{ax^2 - 3a^2 + x}{x^3}$

       (d) $\dfrac{2a + 3b - 4ab}{a^2b^2}$ (e) $\dfrac{6s^3 + 10r^3 - r^2s^2}{4r^2s^2}$

(5)   (a) $\dfrac{a}{3b}$ (b) $3x^3y^3$ (c) $\dfrac{45}{p^2}$ (d) $\dfrac{1}{4(a + b)^3}$

(6)   (a) -25 (b) $\dfrac{3a}{2}$ (c) 51^2 (d) $-\dfrac{y^2}{zx^2}$

       (e) $-\dfrac{2}{k^2l}$

(7)   (a) $x + 2y$ (b) $a - 2$ (c) $3x + 2y$ (d) $\dfrac{3p - 1}{q}$

       (e) $\dfrac{2(a - 1)}{c}$ (f) $q(3q - 2)$ (g) $-\dfrac{(x^2 - 3x + 1)}{2x}$

(8)   (a) $\dfrac{a}{a + b}$ (b) $\dfrac{a - 4}{a - 5}$ (c) $\dfrac{a + 3}{2(a + 4)}$

       (d) $\dfrac{x(x + 2)}{y(x - 3)}$ (e) $\dfrac{x + y}{x - 3}$ (f) $\dfrac{a + b}{2ab}$

       (g) $\dfrac{7y - 3x}{xy}$ (h) $\dfrac{7 - 3x}{xy}$ (i) $\dfrac{7y - 3x}{x^2y^2}$

       (j) $\dfrac{2(2x - 5y)}{(x - y)(x + y)}$ (k) $\dfrac{2(2x + 5y)}{(x - y)(x + y)}$ (l) $\dfrac{2b + 3a}{ab(a - b)}$

       (m) $-\dfrac{5}{(x - 3)(x - 2)(x + 3)}$ (n) $\dfrac{2y}{(x - y)^2(x + y)}$

(9)   (a) $\dfrac{x + 6}{x + 5}$ (b) $\dfrac{ax + 4}{2 - ax}$ (c) 2 (d) $\dfrac{a - b}{a + b}$

(10) (a) $\dfrac{11x}{35}$ (b) $\dfrac{x-15}{(x-3)(x+3)}$ (c) $\dfrac{4-5x}{x(x-3)}$

(d) $\dfrac{5x-21}{(x-3)(x+3)(x+4)}$

(11) (a) $\dfrac{y^2}{2x}$ (b) $\dfrac{x+3}{x-3}$ (c) $\dfrac{x+2}{x-2}$ (d) $\dfrac{x+1}{2}$

(e) $x-y$ (f) $\dfrac{1}{3(x-y)(x+y)}$

(12) (a) $x = \dfrac{a}{a-b}$ (b) $x = 2$

(13) (a) $4a^2$ (b) $\dfrac{16}{a^2}$ (c) 8 (d) $2\left(a^2 + \dfrac{4}{a^2}\right)$

(e) $a^2 - \dfrac{4}{a^2}$ (f) $\dfrac{a^2+2}{a^2-2}$ (g) $\dfrac{2}{a^2}$

Equations and Inequations
simple equations

(1) (a) $x = 5$ (b) $a = 2$ (c) $x = 2$ (d) $b = 7$
(e) $x = 3$ (f) $x = 2$ (g) $x = 1$ (h) $x = \dfrac{13}{5}$
(i) $x = 3$ (j) $a = \dfrac{10}{9}$ (k) $x = 0$ (l) $y = -2$
(m) $a = 4$ (n) $m = -\dfrac{1}{3}$

(2) (a) 5 (b) 7 (c) 9 to 3 (d) 40p

(3) (a) $\{4\}$ (b) $\left\{\dfrac{11}{4}\right\}$ (c) ϕ (d) $\left\{-\dfrac{1}{2}\right\}$
(e) R (f) $\{0\}$ (g) $\left\{\dfrac{5}{2}\right\}$ (h) $\left\{\dfrac{6}{7}\right\}$

(4) (a) $x = -4$ (b) $a = 5$ (c) $y = 3$ (d) $a = 39$
(e) $y = -1$ (f) $x = -9$ (g) $y = -1$ (h) $y = -3$

(5) (a) 4 (b) 11 (c) 5 (d) 8 (e) 10 p/kg (f) 10 p/litre

(6) (a) $x = 24$ (b) $x = -126$ (c) $x = 45$ (d) $x = -3$
(e) $x = 5$ (f) $x = -5$ (g) $y = -4$ (h) $y = 6$
(i) $y = -2\frac{1}{3}$ (j) $a = \dfrac{110}{19}$ (k) $a = 10$ (l) $a = -\dfrac{16}{21}$

simple inequations

(1) (a) $\{1, 2, 3\}$ (b) $\{0, 1, 2\}$ (c) ϕ (d) $\{\frac{1}{2}, 1, 1\frac{1}{2}\}$
(e) $\{y : y \geqslant -1, y \in Z\}$ (f) ϕ
(g) $\{y : y > -3, y \in Z\}$ (h) R

(2) (a) $\{x : x \leqslant 24, x \in R\}$ (b) $\{x : x \geqslant 3\cdot5, x \in R\}$
(c) $\{x : x > 2, x \in R\}$ (d) $\{x : x > -\frac{2}{3}, x \in R\}$
(e) $\{x : -10 < x, x \in R\}$ (f) $\{x : -12 < x, x \in R\}$
(g) $\{x : x < \frac{1}{4}, x \in R\}$ (h) $\{x : x > 1, x \in R\}$
(i) $\{x : x > \dfrac{18}{5}, x \in R\}$ (j) $\{x : -74 \leqslant x, x \in R\}$
(k) $\{x : -\frac{1}{3} \leqslant x, x \in R\}$ (l) $\{x : x < 3, x \in R\}$

(3) (a) $\{x : x < 15, x \in R^+\}$ (c) $\{x : x > \dfrac{7}{5}, x \in R^+\}$

simultaneous equations

(1)   (*a*) $x = 2, y = 1$ (*b*) $x = 3, y = 1$ (*c*) $x = 4, y = 1$
      (*d*) $x = 4, y = -2$ (*e*) $x = -30, y = 40$ (*f*) $x = 8, y = -10$
      (*g*) $x = 6, y = 8$ (*h*) $x = 5, y = 3$ (*i*) $x = 3, y = 5$
      (*j*) $x = -1, y = -1$ (*k*) $x = -2, y = 5$ (*l*) $x = \frac{1}{3}, y = -\frac{5}{6}$

(2)   (*a*) $9\frac{1}{2}$ m, $7\frac{1}{2}$ m   (*b*) 23 of 10p, 7 of 5p   (*c*) $x = 105, y = 25$
      (*d*) 83   (*e*) 53_6 (*f*) 350 of dearer,
                                                            150 of cheaper

quadratic equations

(1)   (*a*) $\{-1{\cdot}6, 0{\cdot}6\}$ (*b*) ϕ (*c*) $\{-3, 1\}$ (*d*) $\{-3{\cdot}2, 1{\cdot}2\}$
      (*e*) $\{3, -1\}$ (*f*) $\{-0{\cdot}6, 3{\cdot}6\}$ (*g*) $\{-2{\cdot}3, -1{\cdot}3\}$

(2)   (*a*) $\{5, 4\}$ (*b*) $\{-13, 2\}$ (*c*) $\{0, 11\}$ (*d*) $\{-6, 6\}$
      (*e*) $\{-4, 2{\cdot}5\}$ (*f*) $\{-1{\cdot}5, 2\}$ (*g*) $\{-\frac{2}{3}, 1{\cdot}5\}$ (*h*) $\{-\frac{1}{4}, 2\}$
      (*i*) $\{-3{\cdot}79, 0{\cdot}79\}$ (*j*) $\{-4{\cdot}20, 1{\cdot}20\}$ (*k*) $\{-3{\cdot}58, -0{\cdot}42\}$
      (*l*) $\{-1{\cdot}47, 1{\cdot}14\}$

(3)   (*a*) $\{-4, -2\}$ (*b*) $\{-1{\cdot}2, 4{\cdot}2\}$ (*c*) $\{0, \frac{2}{3}\}$
      (*d*) $\{-2{\cdot}1, 0{\cdot}79\}$ (*e*) $\{\frac{1}{3}, 4\}$ (*f*) $\{-0{\cdot}78, 1{\cdot}3\}$
      (*g*) $\{-2, 5\}$ (*h*) $\{-4{\cdot}9, 1{\cdot}9\}$ (*i*) $\{0, 5\}$
      (*j*) $\{-4, 4\}$ (*k*) $\{-4{\cdot}4, 3{\cdot}4\}$ (*l*) $\{1{\cdot}4, 1{\cdot}7\}$

(4)   Area is $x(7 - x)$ m². Greatest area is $12\frac{1}{4}$ m², with length and breadth both $3\frac{1}{2}$ m.

(5) $\{2, 4\}$

(6)   (*a*) $2(x - 3)$ cm²; $\frac{3}{2}(x - 2)$ cm²; $2(x - 3)$ cm²; $\frac{3}{2}(x - 2)$ cm²   (*c*) 6

(7)   (*a*) (i) 1   (ii) No   (*b*) $\{-3, 3\}$

(8)   (*a*) $\{-8, 5\}$ (*b*) 15 cm, 5 cm

(9)   (*a*) (i) -8 (ii) 2   (iii) $(-4, 0)$ (*b*) $x = -1$ (*c*) $(-1, -9)$
      (*d*)

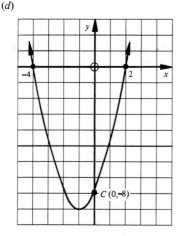

(10)   (*a*) $x(20 - x)$ m²   (*b*) 10   (*c*) 100 m²   (11) $\{0, -6\}$

quadratic inequations

(1) (a) $\{x : x < -2$ and $x > 3\}$ (b) $\{x : x \leqslant -3$ and $x \geqslant 1\}$
      (c) $\{x : -1\tfrac{1}{2} < x < 2\}$ (d) $\{x : x \leqslant 0$ and $x \geqslant 3\}$

(2) (a) $\{x : x \leqslant -4$ and $x \geqslant 10\}$ (b) $\{x : x \leqslant -3$ and $x \geqslant 9\}$
      (c) $\{x : -1 \leqslant x \leqslant 7\}$ (d) $\{x : x \leqslant 0$ and $x \geqslant 6\}$
      (e) $\{x : x \leqslant 2$ and $x \geqslant 4\}$ (f) $\{x : 1 \leqslant x \leqslant 5\}$

(3) $25\,x\,(40 - x) \geqslant 9\,900$; 90 m and 110 m.

(4) (a) (3, 9)
      (b) (i) $\{(x, y) : 6x - x^2 > y \geqslant 0, x < 3 \quad x, y \in R\}$
          (ii) $\{(x, y) : 6x - x^2 > y \geqslant 0, x > 3 \quad x, y \in R\}$
      (c)

Relations, Mappings and Functions

(1) (a) Yes; $f(x) = x + 2$ (b) Yes; $f(x) = x^2 + 1$ (c) No
      (d) Yes; $f(x) = 2x - 5$ (e) Yes   (f) No

(2) (a) Minimum value is -1 when $x = -1$.
      (b) Minimum value is -6 when $y = -1$.
      (c) Minimum value is $-\tfrac{9}{8}$ when $p = -\tfrac{1}{4}$.
      (d) Maximum value is $11\tfrac{1}{2}$ when $x = -\tfrac{1}{2}$.
      (e) Minimum value is $5\tfrac{2}{3}$ when $r = -\tfrac{1}{3}$.

(3)

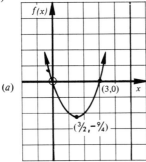

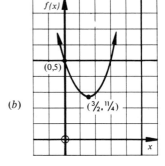

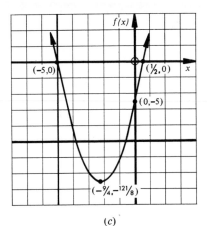

(c)

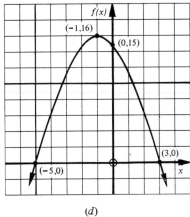

(d)

(4)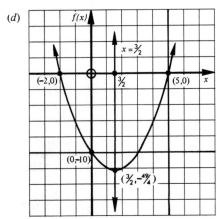

(5) (a) $A = \{1, 2, 3, 4, 5\}$, $B = \{2, 3, 4, 5, 6\}$
(b) $A = \{1, 2, 3, 4, 5, 6, 7\}$, $B = \{1, 2, 3, 4, 5, 6, 7\}$
(c) $A = \{1, 2, 3, 4, 5\}$, $B = \{2, 4, 6, 8, 10\}$
(d) $A = \{x : -1 \leqslant x \leqslant 1, x \in R\}$
$B = \{y : 1 \leqslant y \leqslant 2, y \in R\}$

(6) (i) (a) and (i); (b) and (h); (c) and (l); (d) and (g); (e) and (j); (f) and (k).
(ii) (a), (c) and (d).

(7) (a) $\{-2, 5\}$ (b) $(-2, 0), (5, 0), (0, -10)$

(c) Axis of symmetry, $x = \frac{3}{2}$

(d)

107

(e) Minimum value of $f(x)$ is $-\dfrac{49}{4}$.

(f) Minimum value of $h(x)$ is $-\dfrac{57}{4}$.

(g) Maximum value of $g(x)$ is $+.\dfrac{57}{4}$.

(8) (a) 2   (b) $p = -1, +1, +2$ (c) 0   (d) One

(e) $-1, +1, +2$

(9) (a) (i)

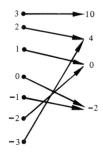

(ii) $\{(-3, 4), (-2, 0), (-1, -2), (0, -2), (1, 0), (2, 4), (3, 10)\}$

(iii)

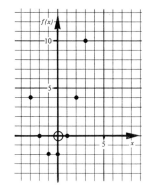

(b) $\{-2, 1\}$

(10) (a) 9   (b) 125

(11) CUMBERLAND;   £A·RR, £C·ED

(12) Range is $\{1, 2, 3\}$;

(13) (a) $\{3, 4, 6, 8, 12\}$
      (b) Range is $\{2, 3, 4, 6\}$;

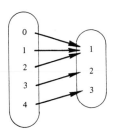

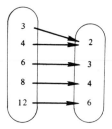

(14) (a) $x = -2$; $(-2, -1)$ is a minimum *TP*.
      (b) $x = -3$; $(-3, 3)$ is a minimum *TP*.
      (c) $x = -2$; $(-2, 12)$ is a maximum *TP*.
      (d) $x = -3$; $(-3, 11)$ is a maximum *TP*.

      (e) $x = -\frac{5}{2}$; $\left(-\frac{5}{2}, -\frac{13}{4}\right)$ is a minimum *TP*.

      (f) $x = -\frac{3}{2}$; $\left(-\frac{3}{2}, \frac{41}{4}\right)$ is a maximum *TP*.

      (g) $x = -\frac{9}{4}$; $\left(-\frac{9}{4}, -\frac{33}{8}\right)$ is a minimum *TP*.

      (h) $x = -\frac{1}{6}$; $\left(-\frac{1}{6}, \frac{97}{12}\right)$ is a maximum *TP*.

(16) (a) 7   (b) ±1·5   (c) ±5   (d) Minimum value of
                                                              $f(x) = -1$.

(17) (a), (b) and (d) are true.

(18) (a) (3, 16)   (b) $(-a, b)$ (c) $\{(-5, 0), (5, 0)\}$ (d) (0, 25)
      (e) $\{x: -5 < x < 5, x \in R\}$

(19) (a) (4, 0)   (b) $x = 2$ (c) 2   (d) −4
      (e) (2, −4)   (f) $0 < x < 4$ (g) Yes   (h) −1, 5
      (i) 12   (j) (6, 12)

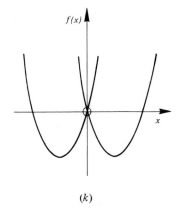

(k)

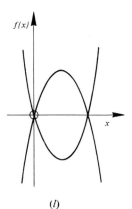

(l)

(20)   (*a*) 14   (*b*) $-5\frac{1}{4}$ (*c*) $(-3, 0); (2, 0)$ (*d*) $(0, -6)$
      (*e*) $-2, 1$

Linear Programming

(1) x of product X and y of product Y
     Raw material: $x + y \leqslant 12$
     Time: $1\frac{1}{2}x + \frac{1}{2}y \leqslant 9$
     For a profit: $4x + 2y$
     Answer: 3 of X, 9 of Y, and a profit of £30.

(2) x of machine A and y of machine B
     Space: $10x + 20y \leqslant 60\,000$
     Weight: $2x + y \leqslant 6\,000$
     For a profit: $50x + 40y$
     Answer: 2 000 of A, 2 000 of B, and maximum profit of £180 000.

(3) x hours of A and y hours of B
     Number of articles: $100x + 150y \leqslant 900$
     Time: $x + y \leqslant 7$
     For a profit: $5\,000x + 6\,000y$
     (*a*) 3 hours of A, 4 hours of B, and a profit of £390.
     (*b*) 5 hours of A, 2 hours of B, and a profit of £390.

(4) x of type A and y of type B
     Length: $10x + 7y \leqslant 378$
     Also: $2y \leqslant x$
     For weight: $10x + 20y$
     Answer: 28 of A, 14 of B, and maximum weight 700 tonnes.

(5) x of small stands and y of large stands
     Area: $4x + 7y \leqslant 308$
     Stands: $y \leqslant x$
     Profit: $30x + 70y$
     Answer: 28 small stands, 28 large stands, and a maximum profit of
              £2 800.

Surds

(1)   (*a*) $2\sqrt{2}$ (*b*) $6\sqrt{2}$ (*c*) $5\sqrt{3}$ (*d*) $9\sqrt{5}$ (*e*) 16
    (*f*) 9   (*g*) 25   (*h*) $3\sqrt{10}$ (*i*) $7\sqrt{10}$ (*j*) 3

    (*k*) 5   (*l*) $\sqrt{\dfrac{7}{3}}$ (*m*) 54   (*n*) 48   (*o*) $45\sqrt{2}$

    (*p*) $16\sqrt{7}$ (*q*) 3   (*r*) $\dfrac{5\sqrt{3}}{8}$ (*s*) 1

    (*t*) $14\sqrt{2}$ (*u*) $37\sqrt{3}$ (*v*) $17\sqrt{5}$

(2)   (*a*) $\dfrac{\sqrt{6}}{3}$ (*b*) $\dfrac{4\sqrt{15}}{5}$ (*c*) $2\sqrt{3}$ (*d*) $\dfrac{3\sqrt{2}}{20}$

(3)   (*a*) $3\sqrt[3]{3}$ (*b*) 4   (*c*) $48\sqrt[3]{2}$ (*d*) 5
    (*e*) $10\sqrt[3]{2} + 12\sqrt[3]{3}$ (*f*) $13\sqrt[3]{3}$

110

Indices

(1)   (a) y^4 (b) c^7 (c) a^8 (d) b^4 (e) 1
       (f) e^{-3} (g) 12   (h) $48y^6$ (i) $0·01x^3$
       (j) $6a^2b^3$ (k) $5^{-1}a^{-6}$ (l) a^{13}

(2)   (a) a (b) y^9 (c) x^4 (d) a^{-2} (e) 64
       (f) $56y^{-7}z^{-8}$ (g) $5a^{-4}$ (h) x^2y^3

(3)   (a) 1   (b) 3   (c) 2   (d) $\dfrac{1}{81}$ (e) $\dfrac{1}{4}$
       (f) $\dfrac{1}{27}$ (g) 12   (h) 6   (i) 54   (j) $\dfrac{1}{6}$
       (k) 108   (l) 216   (m) 24   (n) $\dfrac{1}{6}$
       (o) $\dfrac{243\sqrt[3]{3}}{16}$ (p) $\dfrac{27}{512}$ (q) $\dfrac{3}{2}$

(4)   (a) $\{16\}$ (b) $\{8\}$ (c) $\{-2, 2\}$ (d) $x \in Q,\ x \neq 0$
       (e) ϕ (f) $\{81\}$ (g) $\{4\}$ (h) $\{36\}$
       (i) $\{8\}$ (j) $\left\{\dfrac{3}{2}\right\}$ (k) $\left\{\dfrac{3}{2}\right\}$ (l) $\{4\}$
       (m) ϕ (n) $\{-3\}$ (o) $\{-2\}$ (p) $\{0\}$
       (q) $\{-2\}$

(5) $x = 81, y = 27$

Variation

(1)   (a) 42   (b) 150   (c) 36   (d) 175   (e) $\dfrac{5}{16}$
       (f) 4   (g) 6   (h) 18   (i) 240   (j) $60°$

(2)   199 days   (3) 3460 m

2 GEOMETRY

Triangle

specification, Pythagoras

(2) a, e (3) (a) $a + b > c,\ \ b + c > a,\ \ a + c > b$ (b) (i), (iii) and (iv)
(4)   361 km
(5)   (a) 10 cm   (b) 4 cm   (c) 30 cm   (d) 10·9 cm   (e) 13 cm
       (f) 2·31 cm   (6) 17·5 cm   (7) 6·32 cm
(8)   5 cm, 8·54 cm   (10) 41 cm²
(11)   (a) $\angle ABC$ (b) $\angle PRQ$ (d) $\angle HJI$ (f) $\angle DEF$

similarity

(1)   (c)   (2) (b) and (d)   (3) 5 cm   (4) 64 cm²
(5)   (a) Triangles AED and ABC
       (b) 9 units   (c) 15 units   (d) Angle EDG (e) $\dfrac{2}{3}$ (f) $\dfrac{4}{9}$

(6)   (a) $S = \{\triangle BEA, \triangle CAD, \triangle BDF, \triangle CEF\}; n(S) = 4.$
       (b) (i) 10·4 cm   (ii) 3 cm   (iii) 5 cm   (iv) 5·4 cm   (v) 9 cm
         (vi) 7·2 cm

111

Transformations
reflection, bilateral symmetry
(1) (a) AB', DC, PD' (b) $\angle CAB'$, $\angle CPD$ (c) DS

(2) (a) $P'(-3, -2)$, $Q'(-4, 0)$, $R'(1, 3)$, $S'(0, -1)$, $T'(5, 4)$
  (b) $P'(3, 2)$, $Q'(4, 0)$, $R'(-1, -3)$, $S'(0, 1)$, $T'(-5, -4)$
  (c) $P'(-9, -2)$, $Q'(-10, 0)$, $R'(-5, 3)$, $S'(-6, -1)$, $T'(-1, 4)$
  (d) $P'(3, 0)$, $Q'(4, -2)$, $R'(-1, -5)$, $S'(0, -1)$, $T'(-5, -6)$

(3) (a) Reflection in line $y = -2$ (b) Reflection in line $y = 2$
  (c) Reflection in line $y = x$ (d) Reflection in y-axis
  (e) Reflection in x-axis

(6) (a) Square

translation
(1) (a) $A'(2, 10)$, $B'(-1, 5)$, $C'(-2, 1)$, $D'(6, 11)$
  (b) $A'(-1, 8)$, $B'(-4, 3)$, $C'(-5, -1)$, $D'(3, 9)$
  (c) $A'(1, 1)$, $B'(-2, -4)$, $C'(-3, -8)$, $D'(5, 2)$
  (d) $A'(-1, 4)$, $B'(-4, -1)$, $C'(-5, -5)$, $D'(3, 5)$

(2) (a) $\triangle LRN$ (b) $\triangle JQP$ (c) $\triangle HIO$ (d) $\triangle HGL$
  (e) Half-turn about the mid-point of HL

(3) (a) $\begin{pmatrix} -3 \\ 4 \end{pmatrix}$ (b) $\begin{pmatrix} 3 \\ 0 \end{pmatrix}$ (c) $\begin{pmatrix} -p \\ p-3 \end{pmatrix}$ (d) $\begin{pmatrix} -1 \\ 1 \end{pmatrix}$

rotation, rotational symmetry
(1) (a) (i) $+60°$ (ii) $+180°$ (iii) $+300°$
  (b) (i) C (ii) F (iii) A (iv) CD (v) BD
  (vi) AC (vii) CF (viii) $\angle OFA$ (ix) $\angle BCA$
  (c) Mid-point of OB (d) 0

(2) (a) Intersection of altitudes, $120°$ (b) None
  (c) Intersection of diagonals, $180°$ (d) Intersection of diagonals, $180°$
  (e) Centre (f) Intersection of diagonals, $180°$

(3)

	Image under 90° rotation about origin	Image under 180° rotation	Image under 90° rotation about (2, −1)
(3, 0)	(0, 3)	(−3, 0)	(3, −2)
(−4, 0)	(0, −4)	(4, 0)	(3, 5)
(2, 1)	(−1, 2)	(−2, −1)	(4, −1)
(1, −3)	(3, 1)	(−1, 3)	(0, 0)
(1, −1)	(1, 1)	(−1, 1)	(2, 0)
(3, −2)	(2, 3)	(−3, 2)	(1, −2)
(−1, 3)	(−3, −1)	(1, −3)	(6, 2)

(5) (a) (3, 1) (b) $90°$

(6) (a) (i) (7, 5) (ii) (−2, −1) (b) (i) 1 (ii) 3
  (c) A'', (−7, −5); B'', (2, 1) (d) Parallelogram

dilatations

(2) (a) -1 (b) $(-1, 0)$ (c) $(-1, 0)$
(d) Dilatation, scale factor -1, is equivalent to half-turn about centre of dilatation.
(e) $(0, 1)$ (f) Yes (g) Yes

(3) (a) -1 (b) 2 (c) 4 (d) 3
(e) $\frac{1}{2}$ (f) $\angle ACB$ (g) $\frac{1}{3}$ (h) $[A, -\frac{1}{2}]$

(4) (a) $(0, 0)$ (b) $(-5, 2)$ (c) $(-1, 2)$ (d) $(3, 3)$
(e) $(6, 3)$

(5) (a) 3 (b) 2 (c) -1 (d) $-\frac{1}{2}$ (e) -3

(6) (a) (i) Rectangle $KEGM$ (ii) Rectangle $ADGN$
(b) (i) $[I, -\frac{1}{2}]$ (ii) $[T, -1]$, where T is intersection of
(iii) $[T, -1]$ diagonals of $KJIL$

(7) (a) (i) $\triangle KBD$ (ii) $\triangle JIH$ (iii) $\triangle FDE$ (iv) $\triangle CLJ$ (v) $\triangle GFD$
(b) (i) $[A, 4]$ (ii) $[G, -1]$ (iii) $[L, 2]$
(c) $T_1 = [J, 2]$, $T_2 = [G, -1]$

(8) (a) $(-2, 1)$ (b) $(4, 4)$ (c) $(4, 4)$

Vectors

(1) $\underline{x} = \underline{0}$ (2) (a) 17 (b) $\binom{10}{11}$ (c) $\binom{-4}{-1}$; $\binom{4}{1}$

(3) (a) $\frac{3}{4}(\underline{p} + \underline{q})$ (b) $\frac{1}{4}(3\underline{q} - \underline{p})$

(4) (a) (i) $\binom{4}{-5}$ (ii) $\binom{8}{-10}$ (iii) $\binom{6}{-7\frac{1}{2}}$ (iv) $\binom{-2}{-3}$ (v) $\binom{8}{-4\frac{1}{2}}$
(b) $9 \cdot 2$

(5) (a) $\underline{a} + \frac{1}{3}\underline{b}$ (b) $\frac{1}{2}\underline{a} - \frac{1}{6}\underline{b}$ (c) $\frac{2}{3}\underline{a}$ (d) $\frac{1}{3}\underline{a} - \frac{2}{3}\underline{b}$

(6) (a) (i) $\underline{a}$ (ii) $\underline{c} - \underline{a}$ (iii) $\frac{1}{3}(\underline{c} - \underline{a})$ (iv) $\frac{1}{3}(\underline{c} - \underline{a})$
(v) $\frac{2}{3}\underline{a} + \frac{1}{3}\underline{c}$ (vi) $\frac{2}{3}\underline{a} + \frac{1}{3}\underline{c}$
(b) Parallelogram

(7) (a) $\frac{1}{2}\underline{a}$ (b) $\frac{1}{3}\underline{a}$ (c) $\underline{b} - \frac{1}{2}\underline{a}$ (d) $\frac{1}{3}(\underline{b} - \frac{1}{2}\underline{a})$
(e) $\frac{1}{3}(\underline{a} + \underline{b})$ (f) $\frac{1}{3}\underline{b}$ (g) Parallel, $\frac{DE}{OB} = \frac{1}{3}$

(8) (a) $\overrightarrow{PU}$ represents $\underline{b}$; $\overrightarrow{BC}$ represents $3\underline{b}$, $\triangle ABC$
(b) $k = \frac{1}{3}, l = \frac{1}{3}, m = \frac{1}{3}$ (c) $\underline{x} = \underline{b}, \underline{y} = \underline{c}, \underline{z} = \underline{0}$

Circle, Locus

(1) (a) $(-5, -12)$ (b) 26 (c) $x^2 + y^2 = 169$
(d) $A(13, 0); B(0, 13); C(-13, 0); D(0, -13)$ (e) $\angle APC = 90°$
(f) Square

(2) (b) $(0, 0)$ (c) $(-4, 3)$ (d) (i) $7 \cdot 07$ (ii) $7 \cdot 85$ (iii) $12 \cdot 5$ (iv) $19 \cdot 5$

(3) (a) $\dfrac{180}{\pi}$ (b) $\sqrt{\dfrac{360}{\pi}}$ (c) $\dfrac{90}{\pi}$ (d) 2 (4) $1:1$

(5) (a) $OP = 3$ units (b) $x^2 + y^2 = 9$ (6) $x^2 + y^2 = 4$

(7) Circle, centre $(0, 0)$, radius 4

(8) $\dfrac{x^2}{17^2} + \dfrac{y^2}{15^2} = 1$; $(\pm 17, 0), (0, \pm 15)$

(9) Circle, radius $BC = 10$ cm and centre C

I

Miscellaneous Examples

(1) (*a*) 5   (*b*) (5, 0)   (*c*) 9·49   (*d*) $\begin{pmatrix} 4 \\ -3 \end{pmatrix}$ (*e*) (1, 3)   (*f*) −0·75

(2) (*a*) (6, 2)   (*b*) Parallelogram   (*c*) Dilatation, [*M*, −1]
    (*d*) Kite

(3) (*a*) (i) 3·61   (ii) 7·21   (iii) 8·06
    (*c*) (i) $A \rightarrow C,\ B \rightarrow D$ (ii) $C\,(-3, 2), D\,(-4, -6)$
    (*d*) $E\,(-2, -3); F\,(6, -4); G\,(3, -2); H\,(4, 6)$

(4) (*a*) *C*, (4, 0)   (*b*) Shortest distance between A' and C is straight line.
    (*c*) $6\sqrt{2}$

(5) (*b*) $\frac{1}{3}(\underline{u} - \underline{v})$ (*c*) $\frac{2}{3}(\underline{v} + 2\underline{u})$

(6) (*b*) Half-turn about point (4, 8)   (*c*) Translation $\begin{pmatrix} -8 \\ -16 \end{pmatrix}$ (*d*) (9, 0)

(7) (*b*) Rhombus   (*d*) 1 : 3   (9) 12·07

(10) (*a*) Parallelogram   (*b*) (2, 3); (−1, 4), (−2, −3), (1, −4)
    (*c*) Parallelogram   (*d*) 4 : 1

(12) 6·24 cm

(13) (*a*) $\underline{u} + \frac{3}{4}\underline{v}; \underline{u} - \frac{3}{4}\underline{v}$ (*b*) (i) 2 : 1   (ii) $\frac{2}{3}\underline{u} + \frac{1}{2}\underline{v}$
    (*c*) (i) $\underline{u} + \frac{1}{4}\underline{v}$ (ii) $\underline{u} + \frac{1}{4}\underline{v}$

(14) (*a*) $OA = \sqrt{52} = 2\sqrt{13}; OB = \sqrt{13}; AB = \sqrt{65}$
    (*c*) Isosceles   (*d*) 26   (*e*) 76°

(15) 6·93 cm

(16) (*b*) Rectangle   (*c*) Half-turn about *P*

(17) (*b*) 6 units; 12 units   (*c*) 6 unit²; 24 unit²

(18) (*a*) (i) $\left[A, \dfrac{5}{3} \right]$ (ii) $\left[A, \dfrac{3}{5} \right]$ (*b*) (i) $\left[A, \dfrac{3}{2} \right]$
    (ii) $\left[A, \dfrac{2}{3} \right]$

3 TRIGONOMETRY

Right-angled Triangles

(1) (*a*) $\sin\theta = \dfrac{15}{17}; \cos\theta = \dfrac{8}{17}; \tan\theta = \dfrac{15}{8}$

    (*b*) $\sin\theta = \dfrac{\sqrt{161}}{15}; \cos\theta = \dfrac{8}{15}; \tan\theta = \dfrac{\sqrt{161}}{8}$

    (*c*) $\sin\theta = \dfrac{6}{\sqrt{61}}; \cos\theta = \dfrac{5}{\sqrt{61}}; \tan\theta = \dfrac{6}{5}$

    (*d*) $\sqrt{\dfrac{8}{14}} = \sqrt{\dfrac{4}{7}} = \sin\theta; \cos\theta = \sqrt{\dfrac{3}{7}}; \tan\theta = \sqrt{\dfrac{4}{3}}$

    (*e*) $\sin\theta = \dfrac{\sqrt{545}}{25}; \cos\theta = \dfrac{4\sqrt{5}}{25}; \tan\theta = \sqrt{\dfrac{109}{16}}$

    (*f*) $\sin\theta = \sqrt{\dfrac{17}{24}}; \cos\theta = \sqrt{\dfrac{7}{24}}; \tan\theta = \sqrt{\dfrac{17}{7}}$

(2)   (a) $\angle ACB = 54°$ (b) $AC = 3·53$ cm   (c) $AB = 4·85$ cm

(3)   Range from $35·5°$ to $51·3°$

(4) $\angle DAB = \angle DCB = 96·4°; \angle ADC = \angle ABC = 83·6$

Sine Rule, Cosine Rule, Area

(1) $AC = 16·4$ cm; $BC = 23·5$ cm   (2) $\angle PQR = 51·3°$

(3) $XY = 3·03$ cm; $XP = 2·07$ cm   (4) $\angle GEF = 34·9°$

(5) $BD = 9·21$ cm   (6) $\angle ACB = 104·5°; CD = 3·02$ cm

(7)   (a) $14·7$ cm² (b) 24 cm² (c) $13·4$ cm²   (8) $\angle KLM = 53·1°$ or $126·9°$

(9)   (a) $CA = 38·5$ cm   (b) $\angle ACD = 30°$ (c) Area of $\triangle ACD = 540$ cm²

Three-dimensional Situations

(1)   (a) $AC = 5$ cm   (b) $BG = \sqrt{153}$ cm   (c) $AG = 13$ cm
      (d) Area of $\triangle ABG = 2\sqrt{153}$ cm²   (e) $\angle AGB = 17·9°$

(2)   (a) $DE = 10$ cm   (b) $DF = 26$ cm   (c) $\angle FDE = 67·4°$
      (d) Angle $= 71·6°$

(3)   (a) $TR = \sqrt{153}$ m;   (b) $\angle TRP = 23·8°$ (c) Angle $= 41·5°$

(4)   (a) (i) $DE = 15$ cm   (ii) $AE = 13$ cm   (iii) $DF = 25$ cm
      (iv) $AF = \sqrt{569}$ cm;   (b) $\angle ADF = 68·9°$

(5)   (a) $68·2°$ (b) $60·5°$ (6) (a) $70·6°$ (b) $54·8°$

(7) 96 m   (8) (a) $AM = 3\sqrt{3}$ cm; $BM = 3\sqrt{3}$ cm   (b) $70·6°$

(9)   (a) $\angle AEC = 90°$ (b) $\angle DEC = 90°$ (c) $54·8°$

Bearings

(1)   (a) $AC = 15·5$ km   (b) (i) $\angle DBC = 80°$ (ii) $BD = 15$ km
                       (iii) $DC = 19·3$ km

(2)   (a) $CB = 16$ km;   (b) Distance $= 14·3$ km

(3) $CA = 41·1$ km; $CB = 29·5$ km

(4)   (a) $AC = 39·1$ km; $BS = 71·4$ km; $147·1°$ (b) $19·9$ km/hr

(5)   (a) $(270° \pm 52·9°)$ (b) $41·6°$ or $(90° + 41·6°)$

(6)   Each distance is 65 km

Miscellaneous Examples

(1)   (a) $PQ = 56$ m; $PR = 54$ m   (b) $\angle QPR = 24·9°$
      (c) Area of $\triangle PST = 53$ m²   (d) Area of $STRQ = 582$ m²

(2)   (a) 7 cm (b) 17 cm (c) 13 cm (d) $15·1$ cm (e) $66·4°$ (f) 30 cm²

(3) $60°$ (4) 28 km (approximately)   (5) (a) $x = 120°$ (b) $\frac{1}{2}$

(6)   (a) $x = 8·7$ cm, $y = 5$ cm, $z = 13·2$ cm
      (b) Area of $\triangle ABD = 21·7$ cm², Area of $\triangle EBC = 43·3$ cm²

(7)   (a) $1·08$ km   (b) $AB = 30·0$ km

Appendix — Polar Coordinates

(1) $(4, 30°)$, $(4, 90°)$, $(4, 150°)$, $(4, 180°)$, $(4, 210°)$, $(4, 240°)$, $(4, 270°)$, $(4, 300°)$, $(4, 330°)$

(2) (a) $\sqrt{2}$ (b) $(\sqrt{2}, 45°)$, $(\sqrt{2}, 135°)$, $(\sqrt{2}, 225°)$, $(\sqrt{2}, 315°)$

(3) (b) $(4, 45°)$, $(4, 225°)$ (4) (b) $(3, 135°)$, $(3, 315°)$

(5) (a) (iv)   (b) (iii)

(6) (a) (i) 4 units   (ii) 4 units   (b) $20°$ (c) $10°; 10°$ (d) $40°$
      (e) $(4, 40°)$

(7) (a) $(3, 20°)$ (b) $(6, 100°)$ (c) $(4, 165°)$ (d) $(3, 140°)$
      (e) $(r, 2a° - \theta°)$ (f) $(r, 2a° - \theta°)$ (g) $(r, 2a° - \theta°)$

(8) (b) $(170° - a°)$ (c) $(90° + a°)$

(9) (a) $(2, 30°)$ (b) $(1, 60°)$ (c) $(2, 330°)$ (d) $(6, 225°)$
      (e) $(kr, \theta°)$ (f) $(kr, \theta° + 180°)$ (g) $(kr, \theta° + 180°)$

(10) (a) Rotation of $40°$ (b) Dilatation $[O, 2]$
      (c) Dilatation $[O, -2]$ (d) Dilatation $[O, -\frac{1}{2}]$